E2S2-CREATE and AIChE Waste Management Conference 2019

Singapore
11 – 13 March 2019

ISBN: 978-1-7138-0578-6

-Printed from e-media with permission by:

Curran Associates, Inc.
57 Morehouse Lane
Red Hook, NY 12571

Some format issues inherent in the e-media version may also appear in this print version.

Printed with permission by Curran Associates, Inc. (2021)

For permission requests, please contact AIChE
at the address below.

AIChE
120 Wall Street, FL 23
New York, NY 10005-4020

Phone: (800) 242-4363
Fax: (203) 775-5177

www.aiche.org

Additional copies of this publication are available from:

Curran Associates, Inc.
57 Morehouse Lane
Red Hook, NY 12571 USA
Phone: 845-758-0400
Fax: 845-758-2633
Email: curran@proceedings.com
Web: www.proceedings.com

E2S2-CREATE and AIChE Waste Management Conference 2019

Singapore
11 – 13 March 2019

AIChE
The Global Home of Chemical Engineers

TABLE OF CONTENTS

OPENING CEREMONY AND KEYNOTE SPEECH

CLOSING THE WASTE LOOP: THE ROLE OF RESEARCH AND DEVELOPMENT IN SINGAPORE'S WASTE MANAGEMENT 1
Patrick Pang

KEYNOTE TALK: CIRCULAR ECONOMY WASTE TO RESOURCES MANAGEMENT MINIMISING ENVIRONMENTAL FOOTPRINTS 2
J. Jaromír Klemeš, Yee Van Fan

KEYNOTE TALK: BIOFUELS AND BIOREFINERIES FOR ENERGY AND ENVIRONMENTAL SUSTAINABILITY 3
Ashok Pandey

KEYNOTE TALK: ENVIRONMENTAL SUSTAINABILITY IN WORLD'S FASTEST GROWING REGIONS: CHALLENGES AND OPPORTUNITIES FOR WATER AND WASTE MANAGEMENT 4
Shane Snyder

SECTION A - TOPIC 1, KEYNOTE AND INVITED SPEECH

KEYNOTE TALK: DATA, MACHINE LEARNING AND DECISION-MAKING 5
Jie Lu

INVITED TALK: RESEARCH HIGHLIGHTS ON WASTE PLASTIC 6
Jeffrey R. Seay

INVITED TALK: VALORIZATION OF FOOD WASTE INTO PLATFORM CHEMICALS 7
Daniel C. W. Tsang

INVITED TALK: GREEN AND SUSTAINABLE REMEDIATION USING NATURE BASED SOLUTIONS WITH BIOLOGICAL WASTE DERIVED MATERIALS 8
Deyi Hou

INVITED TALK: BIOMASS PYROLYSIS FOR GAS FUEL, LIQUID CHEMICALS AND SOLID CARBON-MATERIALS POLYGENERATION 9
Haiping Yang, Hanping Chen

INVITED TALK: SOLAR THERMOCHEMICAL TECHNOLOGY FOR PROCESSING CARBONACEOUS MATERIALS 10
Wojciech Lipinski, John Pye, Mahesh Venkataraman, Gilles Maag

INVITED TALK: ENERGY RECOVERY FROM FOOD WASTE GASIFICATION PROCESS USING STIRLING ENGINE WITH HEAT HARVESTING 11
Jialing Chen, Yanjun Dai, Xian Li, Chi-Hwa Wang

SECTION B - TOPIC 2, ORAL PRESENTATION

TIME-DEPEDENT METHANE GENERATION FROM MUNICIPAL SOLID WASTE IN LANDFILLS WITH DIFFERENT MOISTURE CONTENTS AND OVERBURDEN STRESSES 12
Xunchang Fei, Xuhong Lu

CATALYTIC MICROWAVE PYROLYSIS OF PLASTIC WASTE USING METALLIC BIOCHAR DERIVED FROM OIL PALM WASTE: PRODUCTION OF HYDROGEN, GASEOUS HYDROCARBONS AND ENERGY EFFICIENCY 13
Su Shiung Lam

OPTIMAL MAINTENANCE SCHEDULING FOR A WASTE-TO-ENERGY COMBINED HEAT AND POWER PLANT UNDER UNCERTAINTY 14
Chenlian Hu, Xiao Liu, Jie Lu

UPCYCLING OF MIXED PLASTIC WASTE TO OIL AND MULTI-WALLED CARBON NANOTUBES: TECHNO-ECONOMIC FEASIBILITY AND ENVIRONMENTAL IMPACTS 15
Ashiq Ahamed, Andrei Veksha, Ke Yin, Piyarat Weerachanchai, Apostolos Giannis, Grzegorz Lisak

EFFECTS OF PRODUCTION TEMPERATURE ON SUITABILITY OF ENVIRONMENTAL APPLICATIONS OF RICE HUSK BIOCHARS 16
Deyi Hou, Zhengtao Shen, Daniel Alessi

LIFE CYCLE ASSESSMENT OF DIFFERENT STRATEGIES TO OVERCOME AMMONIA INHIBITION IN ANAEROBIC DIGESTION PROCESS 17
Hailin Tian, Xiaonan Wang, Yen Wah Tong

CONCEPTUAL DESIGN OF ETHYLENE OXIDE PRODUCTION PROCESS WITH INTEGRATED CONVERSION OF WASTE CO2 STREAM 18
Magda H. Barecka, Joel W. Ager

SUSTAINABLE RECYCLING OF SPENT ASYMMETRIC-CAPACITANCE POWER BATTERIES 19
Yali Zhang, Xia Chen, Ming Wang, Hongyou Cui

SECTION C - TOPIC 3, KEYNOTE AND ORAL PRESENTATION

KEYNOTE TALK: MEASURING CHINA'S CIRCULAR ECONOMY 20
Yong Geng

INVITED TALK: VERSATILE USE OF CO2 IN THE THERMO-CHEMICAL PROCESS 21
Eilhann Kwon

INVITED TALK: CONTINUOUS THERMAL CONVERSION OF WASTE LIGNIN AND APPLICATION OF LIGNIN DERIVED BIOCHAR FOR THE PRODUCTION OF VALUABLE COMPOUNDS 22
Young-Kwon Park, Se Young Park, Jaehun Jeong, Sumin Ryu

INTEGRATION OF BIOMASS INTO DISTRICT ENERGY SYSTEMS WITH METHANOL PRODUCTION BASED ON A MINLP OPTIMIZATION METHODOLOGY 23
Yingru Zhao, Xiangyan Zhan, Xuyue Zheng, Xingyi Zhu, Jian Lin

UTILIZATION OF WASTE MATERIALS TO ENHANCE THE COMPOSITE BEHAVIOR BETWEEN CEMENTITIOUS MORTAR AND 3D-PRINTED POLYMER REINFORCEMENTS DESIGNED TO BE APPLIED FOR CONSTRUCTION 3D-PRINTING OF CEMENTITIOUS MATERIALS 24
Alexander Lin, Hayden Taylor, Harn Wei Kua, Sze Dai Pang, Jacky Chung, Chi-Hwa Wang

USING AQUATIC PLANTS AND PLANT MATERIAL FOR A SUSTAINABLE WATER PURIFICATION CREATING LOW IMPACT DEVELOPMENT SYSTEMS 25
Stephan Pflugmacher Lima

THE PREDICTION AND OPTIMIZATION OF BIOCHAR YIELD BASED ON THE FEEDSTOCK CHARACTERISTICS AND PYROLYSIS PARAMETERS USING MACHINE LEARNING 26
Xiaonan Wang, Xinzhe Zhu, Yong Sik Ok

GENERATION AND COMBUSTION EMISSIONS OF TRASH TO TANK FUEL DERIVED FROM PLASTIC WASTE .... N/A
Chandni Joshi, Jeffrey R. Seay

PROMOTING FOOD WASTE RECYCLING IN THE COMMERCIAL AND INDUSTRIAL SECTOR BY EXTENDING THE THEORY OF PLANNED BEHAVIOUR: A HONG KONG CASE STUDY 27
Tiffany M. W. Mak, Iris K. M. Yu, Daniel C. W. Tsang, S. C. Hsu, Chi Sun Poon

SECTION D - TOPIC 4, ORAL PRESENTATION

AN ANAEROBIC DIGESTION WASTE-TO-ENERGY SYSTEM FOR FOOD WASTE TREATMENT AND FERTILIZER PRODUCTION 28
Jingxin Zhang, Kai-Chee Loh, Yen Wah Tong

PREPARATION OF NANOCARBON FROM BIOENERGY WASTE DERIVED BIOCHAR THROUGH MECHANIZED GRINDING FOR THE REMOVAL OF PHARMACEUTICALS FROM AQUEOUS MEDIA 29
Sammani Ramanayaka, Yohan Jayawardhana, Heshan Galagedara, Yong Sik Ok, Athula Wijayasinghe, Meththika Vithanage

COMPARISON OF DIFFERENT START-UP STRATEGIES FOR THERMOPHILIC ANAEROBIC DIGESTION: PROCESS STABILITY AND MICROBIAL COMMUNITY STRUCTURE SHIFTS 30
Jun Wei Lim, Yen Wah Tong

TECHNICAL AND ECONOMIC ASPECTS OF HOT SYNGAS PURIFICATION FOR POWER GENERATION FROM GASIFICATION OF MUNICIPAL SOLID WASTE 31
Wei Ping Chan, Andrei Veksha, Xiaomin Dou, Junxi Lei, Kumaran S/o Tamilselvam, Wen-Da Oh, Ze Hua Pan, Apostolos Giannis, Grzegorz Lisak, Teik-Thye Lim

A DATA-DRIVEN TWO-STAGE RESOURCE ALLOCATION MODEL FOR SEWAGE NETWORKS AGAINST FAILURES 32
Jing Jiang, Xiao Liu

CLAY-BIOCHAR NANOCOMPOSITE AS A MEDIA FOR THE REMOVAL OF OXYTETRACYCLINE ANTIBIOTIC FROM WATER 33
Meththika Vithanage, Madara Weerasooriyagedara, Ahmed Ashiq, Binoy Sarkar, Yong Sik Ok

GREEN PREPARATION OF FLUORESCENT CARBON DOTS FROM DURIAN SHELL WASTE FOR HIGHLY SENSITIVE DETECTION OF FE3+ IONS 34
Supuli Jayaweera, Xiao Hu, Ke Yin, Wun Jern Ng

MAKING USE OF WASTED COCONUT WATER THROUGH FREEZE CONCENTRATION........ 35
Daisy Badilla

PINCH ANALYSIS AND EMISSION INTENSITY FOR WASTE MANAGEMENT PLANNING: EU 28 36
Jirí Jaromír Klemes, Yee Van Fan

RESIDENTIAL WASTE DUMPING BEHAVIORS FOR WASTE SORTING AND RELEVANT MANAGERIAL INSIGHTS 37
Peng Jiang, Xiao Liu, Yilu Cai, Minmin Jiang

MICROPLASTICS BOUND TRANSPORT OF ANTIBIOTICS IN AQUATIC ENVIRONMENT 38
Thilakshani Atugoda, Hasintha Wijesekara, Nanthi Bolan, Meththika Vithanage

THREE-STAGE ANAEROBIC CO-DIGESTION OF FOOD WASTE AND WASTE ACTIVATED SLUDGE: IDENTIFYING BACTERIAL AND METHANOGENIC ARCHAEAL COMMUNITIES AND THEIR CORRELATIONS WITH PERFORMANCE PARAMETERS 39
Le Zhang, Kai-Chee Loh, Jingxin Zhang, Liwei Mao

BEHAVIOR OF RECYCLED AGGREGATE REINFORCED CONCRETE COLUMNS UNDER UNIAXIAL LOADING 40
Maham Mustafa, Syed M. Shamaim Ali, Hamais Sajid, Asad-Ur-Rehman Khan, Shamsoon Fareed

SECTION E - TOPIC 5, ORAL PRESENTATION

PHYTOREMEDIATION PLANT 41
Pranjal Mathur, Ayman Hoque, Kishore Kandasamy, Sricharan Poyyamozhi

LIFE-CYCLE ASSESSMENT AND OPTIMIZATION FRAMEWORK FOR SUSTAINABLE URBAN FARMING SYSTEMS........ 42
Lanyu Li, Xian Li, Chi-Hwa Wang, Clive Chong, Xiaonan Wang

MARKET POTENTIAL OF BIOMETHANE IN SOUTH KOREA........ 43
Woong Kim, Ji Hye Jo

PREPARATION OF A NOVEL COMPOSITE USING CHITOSAN, LATERITE AND IRON OXIDE NANOPARTICLES AND ITS EFFICACY IN REMOVING METHYLENE BLUE DYE AND ARSENIC FROM WATER........ 44
Md. Nurus Sakib, Shoeb Ahmed

BATCH AND FIXED BED COLUMN BASED OPTIMIZATION OF MUNICIPAL SOLID WASTE BIOCHAR APPLICATION FOR VOLATILE ORGANIC COMPOUND REMOVAL FORM LANDFILL LEACHATE 45
Yohan Jayawardhana, Meththika Vithanage

THE APPLICATION OF MACHINE LEARNING FOR MODELING THE ADSORPTION OF HEAVY METALS ON BIOCHARS WITH DIFFERENT ORIGINS 46
Xinzhe Zhu, Xiaonan Wang, Yong Sik Ok

ENVIRONMENTAL IMPACTS FROM ENERGY USE IN SOLAR PHOTOVOLTAICS MANUFACTURING 47
Augustine Quek, Alvin Ee

KING COCONUT HUSK DERIVED BIOCHAR THROUGH HYDROTHERMAL ACTIVATION PROCESS FOR THE REMOVAL OF CADMIUM (II) AND COPPER (II) FROM AQUEOUS MEDIA 48
Heshan Galagedara, Meththika Vithanage, Athula Wijayasinghe

POSTER SESSION

APPLICATION OF LIQUID PHASE PLASMA IN PHOTOCATALYTIC HYDROGEN PRODUCTION FROM WASTEWATERS CONTAINING ORGANIC POLLUTANTS 49
Young Kwon Park, Kyong-Hwan Chung, Sang-Chul Jung

CASE STUDY - DEMONSTRATION PLANT OF WATER RECYCLING OF TREATED INDUSTRIAL WASTEWATER USING MICROFILTRATION/REVERSE OSMOSIS AND ACTIVATED CARBON FILTER FOR PROCESS USE IN A PETROCHEMICAL PLANT, SINGAPORE 50
Tan Francis, Boon Siong Ee, Kai Sheng Sim

DEVELOPING METHODOLOGY TO DETERMINE THE EFFICACY OF ODOR REDUCTION TECHNOLOGY OF LIQUID FERTILIZER CIRCULAR OPERATION TECHNIQUE SYSTEM 52
Min-Suk Kim, June H. Hwang, Sang-Ryong Lee

DEVELOPING REDUCTION TECHNOLOGY OF AMMONIUM-SALT BASED INORGANIC SECOND AEROSOL FROM LIVESTOCK MANURE COMPOSTING PROCESS 53
Min-Suk Kim, June H. Hwang, Sang-Ryong Lee, Yong Sik Ok

DEVELOPMENT OF MOS2/GRAPHENE ON CARBON CLOTH ELECTRODE FOR SIMULTANEOUS WASTEWATER TREATMENT AND HYDROGEN PRODUCTION USING WATER SPLITTING 54
Ki-Joon Jeon, Young-Kwon Park

EFFICIENT MG-DOPED POROUS CARBON DERIVED FROM PETROLEUM COKE FOR REMOVING HEAVY METAL IONS FROM WASTEWATER 55
Ki Bong Lee, Xiangzhou Yuan, Min-Jeong Kim

HIGH-THROUGHPUT PROFILING OF ANTIBIOTIC RESISTANT GENES IN A NOVEL TRIDIMENSIONAL ECO-BIOLOGICAL WWTP WITH HIGH DISCHARGE STANDARDS FOR REUSE 56
Kaifeng Yu, Yihan Chen, Bo Zhang, Karina Yew-Hoong Gin, Yiliang He, Peng Li, Yuansheng Huang

INVESTIGATION OF PHOSPHORUS REMOVAL IN SEWAGE SLUDGE SOLID FUEL USING CITRIC AND FORMIC ACID 57
Li-Hua Xu

MESOPHILIC VERSUS THERMOPHILIC ANAEROBIC DIGESTION OF SOYBEAN CURD RESIDUE FOR METHANE PRODUCTION: CHARACTERIZING BACTERIAL AND METHANOGEN COMMUNITIES AND THEIR RELATIONSHIP WITH ORGANIC LOADING RATES 58
Le Zhang, Kai-Chee Loh, Suseeven Sarvanantharajah

PREPARATION OF WASTE COFFEE GROUND-BASED MICROPOROUS CARBONS FOR CO2 CAPTURE 59
Min-Jeong Kim, Ki Bong Lee, Seung Wan Choi

ROBUST TRADING STRATEGIES FOR A WASTE-TO-ENERGY COMBINED HEAT AND POWER PLANT IN A DAY-AHEAD ELECTRICITY MARKET 60
Chenlian Hu, Xiao Liu, Jie Lu

THERMOCHEMICAL CONVERSION OF POLYVINYL CHLORIDE (PVC) IN THE PRESENCE OF FE3O4 AND CO2 FOR SYNGAS PRODUCTION 61
Hocheol Song, Gihoon Kwon, Dong-Wan Cho

UTILIZATION OF GARMENT THERMOSETTING PLASTIC BUTTON RESIDUAL WASTE 62
Md. Iqbal Hossain

Author Index

Closing the Waste Loop: The Role of Research and Development in Singapore's Waste Management

- Conference: E2S2 CREATE and AIChE Waste Management Conference
- Year: 2019
- Proceeding: 2019 E2S2-CREATE and AIChE Waste Management Conference
- Group: Abstract Submissions
- Session:
  Opening Ceremony and Keynote Speech
- Time:
  Tuesday, March 12, 2019 - 9:00am-9:35am

Authors:
Pang, P., National Environment Agency (NEA), Singapore

This presentation will lay out the current waste management landscape in Singapore, focusing on the challenges, which many other cities are facing or would soon face. These will set the context for how the National Environment Agency, along with our partner agencies, are setting about research and development, as well as testbedding and demonstrations, to address these challenges.

Topics:
Sustainability & Environment

Keynote Talk: Circular Economy Waste to Resources Management Minimising Environmental Footprints

- Conference: E2S2 CREATE and AIChE Waste Management Conference
- Year: 2019
- Proceeding: 2019 E2S2-CREATE and AIChE Waste Management Conference
- Group: Abstract Submissions
- Session:
  Opening Ceremony and Keynote Speech
- Time:
  Tuesday, March 12, 2019 - 9:35am-10:10am

Authors:
Klemeš, J. J., Brno University of Technology
Fan, Y. V., Brno University of Technology

Sustainable management considers the efficient resources utilisation and environmental issues for long term gain, benefiting current and future generation. The management practices look beyond the short-term profit. This study aims to review the recent sustainable management strategies and methodologies given focus on environmental components involved in circular economy. The focus is given on the waste/nutrient, water and energy (including renewables) management. The strategies can be generally divided into three major groups: a) Maximise the utilisation of waste as secondary resources b) Minimise the resources used and improve the process efficiency c) GHG and Smog/Haze emission minimisation/mitigation/capture. Circular economy, industrial symbiosis, recycling and Process Integration concepts are the examples of maximising the utilisation of waste. It is mainly by reusing, recovering, sharing (mapping the supply and demand) or even better covering the 6R practice (reduce, reuse, recycle, recover, redesign and remanufacture). The improvement of the process efficiency focuses on upgrading and integrating the technical process by identifying the bottlenecks. GHG minimisation/mitigation/capture for reprocessing is a downstream solution for the released emissions. Several technologies for GHG reprocessing and biochar application are introduced to the circular management system. The GHG capture and storage are more linear and broken economy rather than reuse and circular economy. Environmental impact quantification methods being widely introduced have been environmental footprints. From more than a hundred of developed footprints several of them been analysed: Carbon/GHG Footprint, Nitrogen and Water Footprints, also Virtual Footprints related world-waste trade. The limitation and the need for an extended approach that complements environmental, financial and even societal considerations (e.g. sustainability index) has been also highlighted. This overview is pointing out two important issues in footprint quantification for future research: (i) Afterlife footprints demanding further attention and (ii) Lifespan issues need to be considered.

Topics:
Sustainability & Environment

Keynote Talk: Biofuels and Biorefineries for Energy and Environmental Sustainability

- Conference: E2S2 CREATE and AIChE Waste Management Conference
- Year: 2019
- Proceeding: 2019 E2S2-CREATE and AIChE Waste Management Conference
- Group: Abstract Submissions
- Session:
  Opening Ceremony and Keynote Speech
- Time:
  Tuesday, March 12, 2019 - 10:10am-10:45am

Authors:
Pandey, A., CSIR-Center for Innovation and Translational Research, CSIR-Indian Institute of Technology Research; Center for Energy and Environmental Sustainability

Waste-to-energy concept has gained much momentum in recent years as on one hand, it offers unique opportunity to handle and dispose solid wastes (municipal waste as well as agro-industrial wastes), and simultaneously provides alternative sources of renewable energy. Solid waste treatment and management is a major issue worldwide. Several countries lack proper basic waste management infrastructure and awareness. Thus, waste-to-energy could be an attractive solution for resource recovery, which eventually offers potential benefits when works on principles of biorefinery. A biorefinery is a facility that integrates biomass conversion processes and equipment to produce bio-products, including biofuels and chemicals. It is analogous to today's petroleum refinery. By producing several products, a biorefinery takes the advantages of various components present in the biomass and their intermediates, therefore maximizing the value derived from the biomass feedstock. They also help in complete or near-complete utilization of the feedstock and reduction in solid, liquid or gaseous wastes.

Lignocellulosic biomass-based research has extensively progressed for the production of value-added products and biofuels. Potential application of such biomasses for the production of liquid and gaseous biofuels and other products on principle of biorefineries has gained more attention for possibilities of bioethanol, biobutanol, biodiesel and other high-value chemicals production, coupled with industrial wastewater treatment. Two major pathways for these include thermo-chemical conversion and biochemical conversion. However, process integration is key for the techno-economic success.

The lecture will discuss the issues and perspectives, with specific to Indian scenario.

Topics:
Biorefineries

Keynote Talk: Environmental Sustainability in World's Fastest Growing Regions: Challenges and Opportunities for Water and Waste Management

- Conference: E2S2 CREATE and AIChE Waste Management Conference
- Year: 2019
- Proceeding: 2019 E2S2-CREATE and AIChE Waste Management Conference
- Group: Abstract Submissions
- Session:
  Opening Ceremony and Keynote Speech
- Time:
  Tuesday, March 12, 2019 - 10:45am-11:20am

Authors:
Snyder, S., Nanyang Technological University

Clean and reliable water is critical for sustainable economic development and for protecting human and ecological health. Water is also highly susceptible to impacts of climate change and environmental pollution. Rapid urbanization is taxing natural resources, including water supplies. Before the industrial revolution, cities housed less than 10% of human population. Today, approximately half of the world's population lives in cities, with half of these cities having populations greater than one million people. By 2050, an addition 2.5 billion people are expected to live in cities. Cities produce nearly 70% of greenhouse gas emissions and have a commensure amount of energy consumption. At the same time, highly urbanized areas are increasingly struggling with management of waste generated within these densely populated regions. For instance, synthetic chemicals were not part of the human experience until the late 19th century with the accidental discovery of the synthetic dye, mauve. Today, more than 65 million chemicals are commercially available, many of which enter the environment both knowingly and unknowingly. From a drinking water exposure standpoint, not only are the chemicals themselves a potential risk, but also the innumerable transformation products formed during water treatment processes. Fortunately, technological advances in water purification, waste treatment, and analytical characterization are providing sustainable solutions for protecting the public health and the environment. This presentation will discuss the history and evolution of water and waste treatment technologies with a view for the future opportunities for global sustainability.

Topics:
Sustainability & Environment

Keynote Talk: Data, Machine Learning and Decision-Making

- Conference: E2S2 CREATE and AIChE Waste Management Conference
- Year: 2019
- Proceeding: 2019 E2S2-CREATE and AIChE Waste Management Conference
- Group: Abstract Submissions
- Session:
  Section A - Topic 1, Keynote and Invited Speech
- Time:
  Tuesday, March 12, 2019 - 1:45pm-2:20pm

Authors:
Lu, J., University of Technology, Sydney

The talk will present how machine learning can innovatively and effectively learn from big data to support data-driven decision-making in uncertain and dynamic situations. A set of new fuzzy transfer learning theories, methodologies and algorithms are proposed that transfer knowledge learnt in one or more source domains to target domains by building latent space and mapping functions to overcome tremendous uncertainties in data, learning processes and decision outputs (classification and regression). Another set of concept drift theories, methodologies and algorithms are developed to handle ever-changing dynamic data stream environments with unpredictable stream pattern drifts by effectively and accurately detecting concept drift in an explanatory way, indicating when, where and how concept drift occurs and reacting accordingly. These new developments enable smart learning and therefore enhance data-driven prediction, recommendation and decision support systems in uncertain and dynamic environments.

Topics:
Sustainability & Environment

Invited Talk: Research Highlights on Waste Plastic

- Conference: E2S2 CREATE and AIChE Waste Management Conference
- Year: 2019
- Proceeding: 2019 E2S2-CREATE and AIChE Waste Management Conference
- Group: Abstract Submissions
- Session:
  Section A - Topic 1, Keynote and Invited Speech
- Time:
  Tuesday, March 12, 2019 - 2:20pm-2:40pm

Authors:
Seay, J. R., University of Kentucky

Municipal solid waste accumulation is a major challenge for governmental municipalities, particularly in developing countries, where population growth, urbanization, capital and infrastructure constraints, poor regulation and waste management education challenge the way waste is collected, recycled and disposed. In particular plastic accumulation further poses a threat due to its persistence in the ecosystem when improperly disposed of. This can lead to serious health and environmental consequences when waste plastic is dumped on open plots of land, in the streets or in waterways. Furthermore, plastic accumulation in the world's oceans is a critical sustainability issue. According to some estimates, the mass of waste plastic will outweigh fish by 2050. Unfortunately, much of the plastic that enters the oceans comes from regions where infrastructure limitations are major roadblocks to effective waste management strategies. Finding solutions to address this issue that are environmentally benign, socially acceptable and economically viable will be a critical sustainability challenge. This contribution will present potential strategies and research highlights for addressing this pressing global issue.

Topics:
Sustainability & Environment
Sustainable Engineering

Invited Talk: Valorization of Food Waste into Platform Chemicals

- Conference: E2S2 CREATE and AIChE Waste Management Conference
- Year: 2019
- Proceeding: 2019 E2S2-CREATE and AIChE Waste Management Conference
- Group: Abstract Submissions
- Session:
  Section A - Topic 1, Keynote and Invited Speech
- Time:
  Tuesday, March 12, 2019 - 2:40pm-3:00pm

Authors:
Tsang, D. C. W., Hong Kong Polytechnic Univ.

Food waste that amounts to over one billion tonnes per year globally is a potential renewable feedstock for biorefineries. Our research focuses on the development of high-throughput catalytic systems to produce value-added platform chemicals, such as hydroxymethylfurfural (HMF) and levulinic acid, from various selected food waste that is rich in starch, cellulose, or sugars. Homogeneous (e.g., $SnCl_4$ and $AlCl_3$) and heterogeneous catalysts (e.g., zeolites) were evaluated, which carry different Brønsted acidity and Lewis acidity for controlling HMF yield and selectivity. These acidities maneuver the kinetics of desirable tandem reactions (hydrolysis of glycosidic bonds, isomerization of glucose, and dehydration of fructose) and side reactions (polymerization and rehydration). In particular, our recent studies highlight biochars functionalized via N-doping, metal impregnation, sulfonation, and acid activation as novel solid catalysts in different important reactions, for achieving sustainable and circular biorefineries. Solvents are also evidenced to play an important role beyond serving as a reaction medium, e.g., the production of HMF in the presence of acetone is substantially faster than that in dimethyl sulfoxide/H_2O and tetrahydrofuran/H_2O. Replacing the industrial co-solvent by greener alternatives, i.e., propylene carbonate and Î³-valerolactone, can further accelerate the conversion of food waste. The solvent medium interacts with the substrates and catalysts, altering their reactivity during catalysis. In addition to platform chemicals, hydrochar as a potential solid fuel can be co-generated via a demonstrated microwave-assisted hydrothermal treatment of red seaweed. These research efforts elucidate the roles of different parameters in conversion systems and demonstrate a good potential of food waste valorization for the synthesis of bio-based products in real-life applications.

Topics:
Biorefineries
Design/Green Engineering
Food (Chemicals & Materials)

Invited Talk: Green and Sustainable Remediation Using Nature Based Solutions with Biological Waste Derived Materials

- Conference: E2S2 CREATE and AIChE Waste Management Conference
- Year: 2019
- Proceeding: 2019 E2S2-CREATE and AIChE Waste Management Conference
- Group: Abstract Submissions
- Session:
  Section A - Topic 1, Keynote and Invited Speech
- Time:
  Tuesday, March 12, 2019 - 3:00pm-3:20pm

Authors:
Hou, D., Tsinghua University

Land contamination at many thousands of sites around the world pose a severe hazard to public health. Traditional physical and chemical remediation solutions will often require high energy with secondary pollution emission, high resource input and waste production, and can result in loss of soil functionality. Therefore, the 'Green and Sustainable Remediation' (GSR) concept has been put forward to facilitate prudent holistic land remediation decision making. The emphasis of GSR is placed on maximizing net environmental benefit, rather than health risk assessment alone. This provides a framework where the recycling of waste materials is afforded greater value than the use of virgin materials, and nature based solutions (NBS) are valued over traditional thermal or chemical based remediation. As such, GSR provides the incentive to develop and promote new remediation technologies that are aligned well to these values. This departs from the traditional driving forces for developing new remediation technologies, which has been to develop the best available technology in terms of efficiency of reducing the risk of soil contaminants. In the context of NBS, contaminant stabilization may be achieved by reagents that derive from biological waste, such as biochar. Biochar is produced by the combustion of biomass under a limited supply of oxygen in a controlled environment. The main purpose in contaminated land application is to reduce contaminant bioavailability, as well as enhancing soil structure and fertility. Biochar has the potential be widely used as a low-cost *in situ* stabilization agent for the remediation of pollution-degraded land. The growth of GSR in the remediation industry should proceed hand-in-hand with the development of new nature based solutions like biochar.

Topics:
Biomaterials

Invited Talk: Biomass Pyrolysis for Gas Fuel, Liquid Chemicals and Solid Carbon-Materials Polygeneration

- Conference: E2S2 CREATE and AIChE Waste Management Conference
- Year: 2019
- Proceeding: 2019 E2S2-CREATE and AIChE Waste Management Conference
- Group: Abstract Submissions
- Session:
  Section A - Topic 1, Keynote and Invited Speech
- Time:
  Tuesday, March 12, 2019 - 3:20pm-3:40pm

Authors:
Yang, H., Huazhong University of Science and Technology
Chen, H., Huazhong University of Science and Technology

As the only carbon contained renewable energy resources, the utilization of biomass is the key for the problem between energy and environment. Pyrolysis can convert biomass to gas fuel, liquid oil and solid char efficiently, it can convert biomass to high quality fuel, and is one of the main route for biomass utilization. However, as biomass type is diversity, and pyrolysis is very complex and quickly, the development of biomass pyrolysis is still very slow. In this study, the mechanism of hemicelluloses, cellulose and lignin pyrolysis was revealed. We set up a model to simulate biomass pyrolysis process based on biomass components. Technology of polygeneration based on biomass pyrolysis was put forward with target product controlling theory, and the operation experiment on the pilot and commercial size had also finished. The volume content of combustible composition in bio-gas would be above 70 vol %, especially, the volume content of CH4 would be above 22 vol %, while the LHV of gas would up to 15 MJ/Nm^3. It is a good quality gaseous fuel for user. At the same time, a higher performance on the adsorption and combustion was obtained for the solid char product while the liquid oil was enriched more light weight composition such as phenols and acetic acid. The technology economic and environmental analysis shown that this technology has a good economy and environment performance compared with other biomass conversion technology. A further upgrading of the pyrolytic products and the application of these products were carried out. For example, after nitrogen doping, the char product was converted to a good electrode material with the specific capacitance ~180 F/g. It will be expected as a better biomass conversion technology to enhance the utilization level of abundant biomass resources in China.

Topics:
Biorefineries

Invited Talk: Solar Thermochemical Technology for Processing Carbonaceous Materials

- Conference: E2S2 CREATE and AIChE Waste Management Conference
- Year: 2019
- Proceeding: 2019 E2S2-CREATE and AIChE Waste Management Conference
- Group: Abstract Submissions
- Session:
  Section A - Topic 1, Keynote and Invited Speech
- Time:
  Tuesday, March 12, 2019 - 3:40pm-4:00pm

Authors:
Lipinski, W., The Australian National University
Pye, J., The Australian National University
Venkataraman, M., The Australian National University
Maag, G., Universidade de São Paulo

A review of the state of the art in solar thermal processing of carbonaceous materials is presented. Pertinent studies in process thermodynamics, chemical kinetics, and solar reactor design and modelling are surveyed, and discussed in the context of solar thermal waste processing.

Topics:
Chemical Reaction Engineering
Solar Energy
Sustainability & Environment

Invited Talk: Energy Recovery from Food Waste Gasification Process Using Stirling Engine with Heat Harvesting

- Conference: E2S2 CREATE and AIChE Waste Management Conference
- Year: 2019
- Proceeding: 2019 E2S2-CREATE and AIChE Waste Management Conference
- Group: Abstract Submissions
- Session:
  Section A - Topic 1, Keynote and Invited Speech
- Time:
  Tuesday, March 12, 2019 - 4:00pm-4:20pm

Authors:
Dai, Y., Shanghai Jiaotong University
Chen, J., Shanghai Jiao Tong University
Li, X., National University of Singapore
WANG, C. H., National University of Singapore

The high cost of food waste treatment by the traditional method and the hygiene problems caused by the accumulation make the reduction and energy regeneration of food waste locally necessary. In this paper, a combined gasification and CHP unit using Stirling engine system is modeled and evaluated. The process is aimed to treat the food waste from canteen in the campus of Shanghai Jiao Tong University. After pretreatment, the water content of food waste from the canteens at SJTU is less than 17%. By establishing a steady-state chemical reaction equilibrium model, the characteristics and yield of syngas from food waste through gasification process are analyzed. The results show that CO accounts for about 26.9% and H_2 accounts for about 30.3% in syngas, and the low heating value is 6.674 MJ/m^3. The syngas is fed to a Stirling engine for power generation. The Stirling engine has low requirements of feedstock and is suitable for distributed waste to energy systems. The electricity efficiency and the whole energy efficiency of the Stirling engine is about 27% and 77%. By handling with food waste 1175kg/d during peak days, the system can generate electricity 1594 kW·h/d and output heat in the form of hot water 2953 kW·h/d.

Topics:
Biofuels (Energy)
Energy (Sustainability & Environment)

Time-Depedent Methane Generation from Municipal Solid Waste in Landfills with Different Moisture Contents and Overburden Stresses

- Conference: E2S2 CREATE and AIChE Waste Management Conference
- Year: 2019
- Proceeding: 2019 E2S2-CREATE and AIChE Waste Management Conference
- Group: Abstract Submissions
- Session:
  Section B - Topic 2, Oral Presentation
- Time:
  Tuesday, March 12, 2019 - 1:45pm-2:00pm

Authors:
Fei, X., Nanyang Technological University
Lu, X., Nanyang Technological University

Time-dependent methane (CH_4) generation during biodegradation of municipal solid waste (MSW) in landfills is critical to the design of biogas collection and energy recovery systems. Uncollected CH_4 migrates to the environment and is considered anthropogenic greenhouse gas emission. Therefore, CH_4 generation in landfills is of great interest to environmental engineers and scientists alike. CH_4 generation from MSW concurs with other biochemical, mechanical and hydrological processes in landfills. These processes are coupled and their interactions have not been elucidated clearly. Some of the major factors influencing CH_4 generation are initial waste composition, moisture content (w_c), and overburden stress (σ_v). In this study, the author synthesizes and analyzes the results of more than 20 laboratory large-scale MSW biodegradation experiments available in the literature, among which some experiments were conducted by the authors. Specifically, each time-dependent CH_4 generation curve is characterized by four parameters, CH_4 generation potential (L_0), maximum CH_4 generation rate ($r_{CH4,max}$), lag time before the initiation of CH_4 generation ($t_{rCH4,0}$), and lag time before achieving $r_{CH4,max}$ ($t_{rCH4,max}$). Major findings of the data analysis are: 1) initial waste composition is influential to L_0 and $r_{CH4,max}$; 2) $r_{CH4,max}$ increases with increasing L_0 and both are influenced by w_c; 3) $t_{rCH4,0}$ and $t_{rCH4,max}$ are correlated and both influenced by w_c; and 4) σ_v plays a secondary role in CH_4 generation, although it affects significantly mechanical and hydrological processes of MSW in landfills.

Topics:
Energy (Sustainability & Environment)
Food (Sustainability & Environment)

Catalytic Microwave Pyrolysis of Plastic Waste Using Metallic Biochar Derived from Oil Palm Waste: Production of Hydrogen, Gaseous Hydrocarbons and Energy Efficiency

- Conference: E2S2 CREATE and AIChE Waste Management Conference
- Year: 2019
- Proceeding: 2019 E2S2-CREATE and AIChE Waste Management Conference
- Group: Abstract Submissions
- Session:
  Section B - Topic 2, Oral Presentation
- Time:
  Tuesday, March 12, 2019 - 2:00pm-2:15pm

Authors:
Lam, S. S., University of Cambridge

Microwave pyrolysis was performed on plastic waste pre-mixed with metallic biochar synthesized from similar pyrolysis technique. The metallic biochar, possessing a highly porous structure with high surface area (410 m^2/g), acted initially as 'adsorbent' by providing many active sites for metals, metal oxides, and plastic waste to be attached onto. The biochar then acted as microwave absorbent, which absorbed microwave radiation and heated up to high temperature (650 °C) in a short time, observed to generate arcing and sparks intermittently, and in turn transformed into high temperature hot spots and acted as catalyst that promoted the pyrolysis cracking of plastic waste. This pyrolysis approach provided a fast heating rate (50 °C/min) and short process time (35 min) to generate higher yields of light C_5-C_{10} hydrocarbons (up to 75 wt%), H_2 and CO gases (up to 40 vol%) in the pyrolysis products. The pyrolysis-gas (up to 35 wt%) were dominated by up to 70 vol% of C_1-C_6 hydrocarbons and 40 vol% of H_2 and CO, showing great promise for use as gaseous fuel or to be upgraded to produce more hydrogen as second-generation fuel. Up to 65 wt% yield of liquid oil was also obtained, detected to have high energy content (49 MJ/kg), C_{13}–C_{24} hydrocarbons within the hydrocarbon range of diesel fuel, and promising green features for use as a potentially 'cleaner' fuel with low oxygen and nitrogen content, and free of sulphur. Energy balance analysis showed that this catalytic pyrolysis approach is potentially energy efficient to produce fuel products at a positive energy ratio of 8 (energy content of fuel products/electrical energy supplied for microwave heating) and a net energy output of about 160 MJ/h; this is supported by the high heating rate and short process time recorded by this approach, hence the lower power and energy consumption.

Topics:
Catalysts
Chemical Reaction Engineering
Hydrogen Production & Storage

Optimal Maintenance Scheduling for a Waste-to-Energy Combined Heat and Power Plant Under Uncertainty

- Conference: E2S2 CREATE and AIChE Waste Management Conference
- Year: 2019
- Proceeding: 2019 E2S2-CREATE and AIChE Waste Management Conference
- Group: Abstract Submissions
- Session:
  Section B - Topic 2, Oral Presentation
- Time:
  Tuesday, March 12, 2019 - 2:15pm-2:30pm

Authors:
Hu, C., University of Technology, Sydney
Liu, X., Shanghai Jiao Tong University
Lu, J., University of Technology, Sydney

Waste management and energy crisis are two compelling challenges worldwide due to the population growth, economic development, and rising living standards in both developed and developing countries. Waste-to-energy (WtE) technologies, which turn waste into energy, provide elegant and effective solutions to these challenges. Because of the high overall thermal efficiency, modern WtE plants are often combined heat and power (CHP) producers which generate power and heat simultaneously. WtE CHP plants are often integrated into a district heating network and they also sell electricity on commercial terms. To ensure the continuous operation of WtE CHP plants, preventive maintenance scheduling is a significantly important task for plant operators. This paper studies the optimal preventive maintenance scheduling strategy for a WtE CHP plant which exports heat to a district heating network and sells electricity in a deregulated electricity market. Since the maintenance scheduling strategy has to be determined months before the actual actions, many key parameters are uncertain or partly predictable at the time of planning the maintenance. These parameters include future electricity prices, district heating demand, and the amount of waste delivered to the plant. To address the problem, we develop a two-stage robust optimization model considering different types of uncertainty. We also design a solution procedure based on the column-and-constraint generation method to solve the model. To illustrate the effectiveness of the proposed model and the solution procedure, a case study is performed based on real-world data. Moreover, a sensitivity analysis is conducted to investigate the impacts of different types of uncertainty on the optimal maintenance scheduling strategies for the WtE CHP plant.

Topics:
Energy Production
Plant Operations

Upcycling of Mixed Plastic Waste to Oil and Multi-Walled Carbon Nanotubes: Techno-Economic Feasibility and Environmental Impacts

- Conference: E2S2 CREATE and AIChE Waste Management Conference
- Year: 2019
- Proceeding: 2019 E2S2-CREATE and AIChE Waste Management Conference
- Group: Abstract Submissions
- Session:
  Section B - Topic 2, Oral Presentation
- Time:
  Tuesday, March 12, 2019 - 2:30pm-2:45pm

Authors:
Ahamed, A., Nanyang Technological University
Veksha, A., Nanyang Technological University
Yin, K., Nanjing Forestry University
Weerachanchai, P., Nanyang Technological University
Giannis, A., Nanyang Technological University
Lisak, G., Nanyang Technological University

Large amounts of mixed plastic waste are generated worldwide, including in Singapore. The available technologies, such as mechanical sorting, chemical treatment and pyrolysis of mixed plastics to fuels have low economic benefits. Owing to low profitability, the recycling rates for mixed plastic waste are low. One of the possible solutions is the conversion of mixed plastics into carbon nanotubes (CNTs) that have high value (> 80 USD/kg) and thus, create an additional stream of revenue for pyrolysis plant operators encouraging waste recycling. The objectives of this study were to investigate the techno-economic feasibility and environmental impacts of the integrated process that upcycles mixed plastic waste with high PET content (11.8 and 27.5 wt.% of PET) into oil and CNTs. In the proposed process, mixed plastics were initially pyrolyzed and the produced vapours were subjected to catalytic treatment to improve oil quality. After oil condensation, the remaining non-condensable pyrolysis gases were used for CNTs synthesis. Cost-benefit analysis indicated 3-10 times more revenue for the plant operators with the addition of CNTs synthesis unit. The life cycle assessment (LCA) study indicated that fossil depletion and climate change potentials for the feedstock with 27.5 wt.% PET are 30% and 73% higher, respectively, compared to the feedstock with 11.8 wt.% PET. Integration of CNTs synthesis with the pyrolysis process showed significant improvements in the climate change, fossil depletion, eco-toxicity and acidification potentials yielding positive effects on the environment. Overall, the pyrolysis process integrated with CNTs synthesis showed environmental benefits while incineration resulted in net negative impacts.

Topics:
Business Strategy and Economics (Sustainability & Environment)
Environmental Sustainability
Nanomaterials

Effects of Production Temperature on Suitability of Environmental Applications of Rice Husk Biochars

- Conference: E2S2 CREATE and AIChE Waste Management Conference
- Year: 2019
- Proceeding: 2019 E2S2-CREATE and AIChE Waste Management Conference
- Group: Abstract Submissions
- Session:
  Section B - Topic 2, Oral Presentation
- Time:
  Tuesday, March 12, 2019 - 2:45pm-3:00pm

Authors:
Shen, Z., University of Alberta
Hou, D., Tsinghua University
Alessi, D., University of Alberta

The environmental application of biochar in soil or water treatment depends on the surface chemistry of biochar and its interaction mechanisms with heavy metals. In this study, rice husk biochars were produced at 300, 500, and 700 °C (RH300, RH500, and RH700). The surface chemistry of these rice husk biochars and their interaction mechanisms with lead (Pb) were investigated. The surface area (from 0.632 to 193.149 m^2/g) and pH (from 7.13 to 9.80) of the rice husk biochars significantly increased as production temperature rose from 300 to 700 °C, while the number of functional groups (e.g., carboxyl) decreased. The maximum adsorption capacity (Q_{max}) values for Pb are in the order of RH300<RH500<RH700 (14.1, 21.7, and 26.7 mg/g respectively). Although RH300 has the smallest Q_{max} value, its exchangeable Pb amount is the largest (2.61 versus 0.223-0.377 mg/g), suggesting RH300 may be suitable for water treatment due to the easy separation of immobilized Pb and better recycling usage. The Pb immobilized on RH500 and RH700 was mainly acidic soluble and generally stable. Hydrocerussite ($Pb_3(CO_3)_2(OH)_2$) is one important form within the acidic soluble fraction. Within the generally stable formation, pyromorphite ($Pb_5(PO_4)_3X$ (X=Cl, F, OH)) is a form for the immobilized Pb on the rice husk biochars, particularly for RH500 and RH700. These findings suggest RH500 and RH700 are of huge potential to be applied in soil remediation to immobilize Pb and reduce its environmental risks.

Topics:
Biomaterials

Life Cycle Assessment of Different Strategies to Overcome Ammonia Inhibition in Anaerobic Digestion Process

- Conference: E2S2 CREATE and AIChE Waste Management Conference
- Year: 2019
- Proceeding: 2019 E2S2-CREATE and AIChE Waste Management Conference
- Group: Abstract Submissions
- Session:
  Section B - Topic 2, Oral Presentation
- Time:
  Tuesday, March 12, 2019 - 3:00pm-3:15pm

Authors:
Tian, H., National University of Singapore
Wang, X., National University of Singapore
Tong, Y. W., National University of Singapore

Anaerobic digestion (AD) is a widely used technology to simultaneously dispose biowaste and recover energy (in the form of CH_4). However, the hydrolysis of urea (e.g. manure-based feedstock) and the degradation of protein and amino acid releases ammonia, which could inhibit the AD process when it is at high levels (> 0.15 g NH_3-N L^{-1}). Thus poor substrate utilization and unstable reactor performance are usually observed when protein-rich feedstock is used. In recent years, different strategies are proposed to overcome the ammonia inhibition problem, aiming to improve the methane yield of the feedstock. For example, bioaugmentation with ammonia tolerant methanogenic consortia, dilution of the substrate with water, addition of absorbing material and air stripping are extensively investigated by a lot of researchers. These methods increase the methane production to a certain level, but they also need extra operations and costs, which cause additional environmental impacts, such as energy and resource consumption. Up to now, few studies can be found discussing about the overall environmental impacts of these proposed methods. Therefore, in order to determine which method to use from the sustainable perspective, it is of great importance to evaluate and compare the environmental impacts of all the methods. In this study, life cycle sustainability assessment (LCSA) methodology is employed to assess these methods with respect to both the life cycle costing and the environmental impacts on different damage categories, such as climate change, human heath, resource, etc. The results derived from this study will be used to identify the weak point of each method and support the strategy choosing to overcome ammonia inhibition in the future.

Topics:
Case Studies (Sustainability & Environment)
Energy (Sustainability & Environment)
Environmental Sustainability

Conceptual Design of Ethylene Oxide Production Process with Integrated Conversion of Waste CO2 Stream

- Conference: E2S2 CREATE and AIChE Waste Management Conference
- Year: 2019
- Proceeding: 2019 E2S2-CREATE and AIChE Waste Management Conference
- Group: Abstract Submissions
- Session:
  Section B - Topic 2, Oral Presentation
- Time:
  Tuesday, March 12, 2019 - 3:15pm-3:30pm

Authors:
Barecka, M. H., TU Dortmund University
Ager, J. W. III, Joint Center for Artificial Photosynthesis

Carbon dioxide (CO2) is widely known to be one of the most redundant gaseous waste streams. The conversion of CO2 into fuels and chemicals is a promising pathway for dealing with waste CO2 and more sustainable chemicals manufacturing. Nevertheless, examples of the process on an industrially relevant scale are still missing. One of the obstacles limiting the process application are additional up & downstream costs related to obtaining pure CO2 and separation of the reaction products. Hence, this paper analyses the case of integration of CO2 electroreduction (CO2RR) into a manufacturing process in which those two key barriers can be omitted: ethylene oxide (EO) production. EO is one of most widely used bulk chemicals and the process variant mostly applied in the industry generates significant amounts of waste CO2 as reaction by-product. The waste CO2 is separated from gas stream and removed to avoid compound build-up, hence the substrate for electroreduction is available at no additional cost. CO2 can be converted into an ethylene-rich stream and recycled back to the EO reactor, which uses ethylene as raw material. Further, no complex separations are required, as other key electroreduction product (CH4) is also used in EO production reactors. Overall, integration of CO2 electroreduction into EO manufacturing enables to reduce waste generation and significantly increase process mass efficiency. To highlight the great potential of such process integration, the conceptual process design is presented and the potential for economic savings is quantified.

Topics:
Carbon Management
Process Design & Development

Sustainable Recycling of Spent Asymmetric-Capacitance Power Batteries

- Conference: E2S2 CREATE and AIChE Waste Management Conference
- Year: 2019
- Proceeding: 2019 E2S2-CREATE and AIChE Waste Management Conference
- Group: Abstract Submissions
- Session:
  Section B - Topic 2, Oral Presentation
- Time:
  Tuesday, March 12, 2019 - 3:30pm-3:45pm

Authors:
Wang, M., Shandong University of Technology
Zhang, Y., Shandong University of Technology
Chen, X., Shandong University of Technology
Cui, H., Shandong University of Technology

There is a pressing need to recover asymmetric-capacitance nickel-metal hydride power batteries (AC-NiMHs) owing to their large number of applications. However, recycling of these batteries is challenging since it is a long process with low extraction efficiency and high cost. Herein, we demonstrate a sustainable, green, and effective recovery process for rare earth metals while directly preparing $LiNi_{0.6}Co_{0.2}Mn_{0.2}O_2$ cathodes from cathode and anode AC-NiMH spent materials. We investigate the effects of a range of parameters (i.e., acid concentration, reaction temperature, and time) on the leaching process. Over 99% of rare earth, Co, Ni, and Mn are leached out during the recovery process, with the leaching kinetics following a shrinking core model with inter-diffusion as the rate-determining step. Rare earths are recovered in the form of high-purity sulfuric acid complex salts, while Ni, Co, and Mn are used to synthetize $LiNi_{0.6}Co_{0.2}Mn_{0.2}O_2$ cathodes with good electrochemical properties. This research demonstrates a sustainable process for the recycling of power sources by incorporating the principles of green chemistry.

Topics:
Environmental Sustainability

Keynote Talk: Measuring China's Circular Economy

- Conference: E2S2 CREATE and AIChE Waste Management Conference
- Year: 2019
- Proceeding: 2019 E2S2-CREATE and AIChE Waste Management Conference
- Group: Abstract Submissions
- Session:
  Section C - Topic 3, Keynote and Oral Presentation
- Time:
  Wednesday, March 13, 2019 - 9:00am-9:35am

Authors:
Geng, Y., Shanghai Jiao Tong University

China is pioneering a new sustainable development model which has the ability to overcome current environmental and resource management problems, while achieving improvements in resource productivity and eco-efficiency. This model, formally accepted in 2002 and termed the "circular economy", is understood to mean the realization of a closed-loop of materials flow in the Chinese economic system. Successful implementation of this model is seen as one way China can "leapfrog" the environmental damages typically seen as economies industrialize. This speech introduces an emergy based circular economy indicators, including the objectives, policies and measures, so that the country can 'leapfrog' its way from environmentally-damaging development to a more sustainable path. The speech then identifies a series of barriers and challenges to the implementation of the emergy-based indicators In China.

Topics:
Sustainable Supply Chains

Invited Talk: Versatile Use of CO2 in the Thermo-Chemical Process

- Conference: E2S2 CREATE and AIChE Waste Management Conference
- Year: 2019
- Proceeding: 2019 E2S2-CREATE and AIChE Waste Management Conference
- Group: Abstract Submissions
- Session:
  Section C - Topic 3, Keynote and Oral Presentation
- Time:
  Wednesday, March 13, 2019 - 9:35am-9:55am

Authors:
Kwon, E., Sejong University

Valorization of biomass and municipal solid waste (MSW) via a pyrolytic platform was investigated. In an effort to establish the more sustainable pyrolysis platform for biomass and MSW, this study particularly employed carbon dioxide (CO_2) as reactive gas medium. Thus, this study laid great emphasis on mechanistic understanding of CO_2 during the thermolysis of biomass and MSW. A series of the TGA tests signified that the mechanistic roles of CO_2could be exerted by the homogeneous reactions between volatile organic carbons (VOCs) evolved from the thermolysis of carbonaceous materials (*i.e.,*biomass and MSW) and CO_2. Also, lab-scale pyrolysis of biomass and MSW demonstrated that the homogeneous reactions (between VOCs and CO_2) played a pivotal role for shifting the carbon distribution from pyrolytic oil to pyrolysis gas. Such phenomena were initiated at = 550 °C. Meanwhile, CO_2also suppressed dehydrogenation of VOCs. Such genuine mechanistic role of CO_2resulted in the compositional modification of pyrolytic oil by restricting the formation of benzene derivatives including polycyclic aromatic hydrocarbons (PAHs).

Topics:
Carbon Capture & Storage
Carbon Management
Energy (Sustainability & Environment)

Invited Talk: Continuous Thermal Conversion of Waste Lignin and Application of Lignin Derived Biochar for the Production of Valuable Compounds

- Conference: E2S2 CREATE and AIChE Waste Management Conference
- Year: 2019
- Proceeding: 2019 E2S2-CREATE and AIChE Waste Management Conference
- Group: Abstract Submissions
- Session: Section C - Topic 3, Keynote and Oral Presentation
- Time: Wednesday, March 13, 2019 - 9:55am-10:15am

Authors:
Park, Y. K., University of Seoul
Park, S. Y., University of Seoul
Jeong, J., University of Seoul
Ryu, S., University of Seoul

Recently, large amount of waste lignin has been produced from the bioalcohol (ethanol, butanol) synthesis process using lignocelluosic biomass as feed materials. Also, many researches have been performed for the conversion of waste lignin to valuable monomeric phenolics using fast pyrolysis process to increase the economic feasibility of bioalcohol production process. However, lignin is difficult to be converted effectively to liquid fuels because of its complex structure and thermal stability. Especially, melting of lignin and foaming and agglomeration of char during pyrolysis make a continuous lignin pyrolysis process difficult in a conventional fluidized reactor. Therefore, the foaming and agglomeration of lignin char needs to be minimized by developing a *new reactor*.

In this study, fast pyrolysis of lignin was carried out using a newly developed bench scale rotary kiln reactor. Initially alumina ball was filled in the reactor. Lignin was then fed into the rotary kiln reactor with a feeding rate of 100g/h. Various kinds of lignin such as kraft lignin, organosolv lignin, butanol production waste lignin, and black liquor lignin were tested. By using this new rotary kiln reactor, the agglomeration and foaming of char did not occurr. Also, highly valuable monomeric phenolics were obtained as the main products. Furthermore, lignin-derived biochar with Ni impregnation was applied to the gasification of biomass and was found to produce higher amount of hydrogen gas than commercial Ni/alumina catalyst.

This work was supported by the New and Renewable Energy Core Technology Program of the Korea Institute of Energy Technology Evaluation and Planning, granted financial resource from the Ministry of Trade, Industry & Energy, Republic of Korea (No. 20173010092430). Also, this work was supported by the National Research Council of Science & Technology (NST) grant by the Korea government (MSIP) No. CAP-16-05-KIMM.

Topics:
Biofuels (Energy)
Biorefineries
Energy (Sustainability & Environment)

Integration of Biomass into District Energy Systems with Methanol Production Based on a MINLP Optimization Methodology

- Conference: E2S2 CREATE and AIChE Waste Management Conference
- Year: 2019
- Proceeding: 2019 E2S2-CREATE and AIChE Waste Management Conference
- Group: Abstract Submissions
- Session:
  Section C - Topic 3, Keynote and Oral Presentation
- Time:
  Wednesday, March 13, 2019 - 10:15am-10:30am

Authors:
Zhao, Y., Xiamen University
Zhan, X., Xiamen University
Zheng, X., Xiamen University
Zhu, X., Xiamen University
Lin, J., Xiamen University

The international energy consumption has considerably increased due to the demographic and economic growth in recent decades. In this framework, polygeneration represents an innovative and promising concept. The purpose of this study is to provide a generic modelling and optimization framework to evaluate the polygeneration systems in urban areas. An integrated biomass-based polygeneration scheme with CCHP (combined cooling, heating and power) system, providing electricity, heating and cooling as well as methanol product, is proposed and evaluated in this work. The potential of integrating the system into a district energy system is also investigated by case study. A mixed-integer nonlinear programming (MINLP) approach has been formulated in GAMS (General Algebraic Modeling System) to address the following issues: a) selection of technologies, types and numbers of equipment to match the energy demand of end users; b) strategic planning and process designing from both economic and environmental perspectives; c) representation of the relationships and trade-offs between the trigeneration and chemical conversion steps; d) operation strategies for the dynamic energy systems with time dependent components and boundary conditions. Moreover, the proposed system is compared to the conventional natural gas based CCHP system as the reference.

Topics:
Bioenergy
Energy (Sustainability & Environment)
Energy Efficiency

Utilization of Waste Materials to Enhance the Composite Behavior between Cementitious Mortar and 3D-Printed Polymer Reinforcements Designed to be Applied for Construction 3D-Printing of Cementitious Materials

- Conference: E2S2 CREATE and AIChE Waste Management Conference
- Year: 2019
- Proceeding: 2019 E2S2-CREATE and AIChE Waste Management Conference
- Group: Abstract Submissions
- Session:
  Section C - Topic 3, Keynote and Oral Presentation
- Time:
  Wednesday, March 13, 2019 - 10:30am-10:45am

Authors:
Lin, A., National University of Singapore
Taylor, H., University of California, Berkeley
Kua, H. W., National University of Singapore
Pang, S. D., National University of Singapore
Chung, J., National University of Singapore
Wang, C. H., National University of Singapore

Polymer reinforcement with novel geometry achieved by 3D-printing was proposed as a new reinforcement system for 3D-printed mortar components, which consist of cement, water and silica sands, for buildings and infrastructures. This polymer reinforcement effectively enhanced the ductility of mortar specimens which was casted layer by layer to simulate a mortar 3D-printing process. However, the polymer reinforcement system reduced the mechanical strength of the mortar specimens. Waste materials, biochar and fly ash, were added to mortar not only to reduce the carbon footprint but also to find potential solutions for improving the mechanical behavior of the mortar-polymer composite because these waste materials can potentially densify the mortar-polymer interface which is essential for composite behavior between mortar and polymer. Experimental results showed that using biochar, a waste material made by sawdust, in mortar specimens can reduce the mechanical strength reduction induced by polymer reinforcement compared to the strength reduction observed in the mortar without waste material.

Topics:
Materials

Using Aquatic Plants and Plant Material for a Sustainable Water Purification Creating Low Impact Development Systems

- Conference: E2S2 CREATE and AIChE Waste Management Conference
- Year: 2019
- Proceeding: 2019 E2S2-CREATE and AIChE Waste Management Conference
- Group: Abstract Submissions
- Session:
  Section C - Topic 3, Keynote and Oral Presentation
- Time:
  Wednesday, March 13, 2019 - 10:45am-11:00am

Authors:
Pflugmacher Lima, S., University of Helsinki

Phytoremediation and the use of living plants to clean up soil, air or water from hazardous contaminants is well known as a cost-effective approach to take advantage of the abilities of plants to take up compounds and metabolize them in their tissues. However, the practical use of plants and associated microorganisms can lead to an external metabolization of compounds giving possibilities in generating even more toxic metabolites. Recently we demonstrated the successful use of systems only containing rootless aquatic plants for the remediation of cyanobacterial toxins, as well as veterinary pharmaceuticals such as oxytetracycline and methyltestosterone. These Low Impact Development Systems, we called “Green Liver Systems” in analogy that plant have the same or very similar biotransformation systems like our own human liver. Our research showed that using the natural functions aquatic macrophytes can provide in taking up and metabolizing various contaminants from water is beneficila to the environment. The Green Liver Concept showed that the metabolites are than stored within the plant in vacuols, apoblasts and also included in cell walls. In order to make Green Liver System multi-functionable and suitable for the removal of microplastics as well, we developed additional filter systems using dead plant material such as coconut fibers, luffa pads, jute fibers, cotton pads derived from used t-shirts having a recycling idea in mind and biochar derived from Finnish spruce trees. Applying microplastics (spherical beads 2-10mm, cylindrical particles (1-5mm) and fleece fibers) on the different plant material showed a retention between 75 % - 96 % (spherical beads), 100 % zylindrical particles and 100 % fleece fibers. The overall performance of the newly developed LID System was tested using Diclofenac in combination with spherical microbeads (microplastics) showing that 85% of the Diclofenac could be removed from the water phase and 99% of the MP as well.

Topics:
Environmental Sustainability
Water (Sustainability & Environment)

The Prediction and Optimization of Biochar Yield Based on the Feedstock Characteristics and Pyrolysis Parameters Using Machine Learning

- Conference: E2S2 CREATE and AIChE Waste Management Conference
- Year: 2019
- Proceeding: 2019 E2S2-CREATE and AIChE Waste Management Conference
- Group: Abstract Submissions
- Session:
  Section C - Topic 3, Keynote and Oral Presentation
- Time:
  Wednesday, March 13, 2019 - 11:00am-11:15am

Authors:
Wang, X., National University of Singapore
Zhu, X., National University of Singapore
Ok, Y. S., Korea Biochar Research Center, O-Jeong Eco-Resilience Institute (OJERI) & Division of Environmental Science & Ecological Engineering, Korea University

Lignocellulosic biomass, as one of the most abundant and promising renewable raw materials producing biofuels, have attracted extensive attention for the dual pressures from increasing energy demand and environmental pollution caused by fossil fuel. The feedstock was derived from the living or waste plants such as woody and agricultural residue. Biochar was the solid by-product of biomass conversion, which has been applied in multidisciplinary due to the higher surface area, microporous structures, and containing functional groups. Generally, the functional groups on the biochar surface area decreased and the surface area increased along with the pyrolysis temperature. Biochar with different characteristics could be applied for different fields. For example, the higher surface area and microporous structure was the most important factor for CO2 capture, while the adsorption of biochar for heavy metals was significantly influenced by the functional groups and containing mineral. Accordingly, improving the biochar yield was one of the important research fields based on biomass properties. Machine learning method was used in the study to predict the biochar yield based on the structural and elementary components of biomass, the particle size of biomass and the pyrolysis parameters. The relative contribution of the influencing factors to biochar yield was also estimated. The results demonstrated the influence of pyrolysis temperature was overwhelming for biochar yield. The partial dependence plots were also performed to gain insights into the relationship between biochar yield and each feature or the synergy effect of any two features on the biochar yield, which could guide the selection of pyrolysis parameters based on different biomass properties.

Topics:
Carbon Management
Thermodynamics

Promoting Food Waste Recycling in the Commercial and Industrial Sector By Extending the Theory of Planned Behaviour: a Hong Kong Case Study

- Conference: E2S2 CREATE and AIChE Waste Management Conference
- Year: 2019
- Proceeding: 2019 E2S2-CREATE and AIChE Waste Management Conference
- Group: Abstract Submissions
- Session: Section C - Topic 3, Keynote and Oral Presentation
- Time: Wednesday, March 13, 2019 - 11:30am-11:45am

Authors:
Mak, T. M. W., The Hong Kong Polytechnic University
Yu, I. K. M., University of York
Tsang, D. C. W., Hong Kong Polytechnic Univ.
Hsu, S. C., The Hong Kong Polytechnic University
Poon, C. S., The Hong Kong Polytechnic University

In recent years, the government and community have focused extensive attention on food waste recycling to achieve sustainable waste management. However, the motivation of stakeholders from the commercial and industrial sector to change recycling behaviour has not received enough attention. This study aimed to identify, prioritise, and quantify the relationships between key latent variables that affect the food waste recycling behaviour of relevant industries in Hong Kong (*i.e.* representatives from the hotel, food and beverages, and property management industries). This study integrated semi-structured interviews and a survey questionnaire on the basis of the Theory of Planned Behaviour, setting it apart from conventional studies on food waste behavioural factors. Qualitative content analysis and quantitative structural equation modelling were performed to analyse the interview and questionnaire responses, followed by correlation analysis to quantify the relationships between variables. The results demonstrate that food waste recycling behaviour is determined by three latent variables—administrative incentives and corporate support, logistics and management incentives, and economic incentives. Administrative incentives and corporate support demonstrate significant effects on recycling behaviour. Moral attitudes show a strong positive correlation with administrative incentives and corporate support. Similarly, there is a significant positive correlation between moral attitudes and logistics and management incentives. Administrative incentives and corporate support is the determining variable for the hotel and food and beverages industries, whereas logistics and management incentives are of the highest concern to property management representatives. These findings can facilitate the development of stakeholder-oriented policy to encourage corporate behavioural change towards food waste recycling for sustainable resource circulation.

Topics:
Food (Sustainability & Environment)
Social Aspects (Sustainability & Environment)
Young Professionals

An Anaerobic Digestion Waste-to-Energy System for Food Waste Treatment and Fertilizer Production

- Conference: E2S2 CREATE and AIChE Waste Management Conference
- Year: 2019
- Proceeding: 2019 E2S2-CREATE and AIChE Waste Management Conference
- Group: Abstract Submissions
- Session:
  Section D - Topic 4, Oral Presentation
- Time:
  Wednesday, March 13, 2019 - 9:00am-9:15am

Authors:
Zhang, J., Shanghai Jiao Tong University
Loh, K. C., National University of Singapore
Tong, Y. W., National University of Singapore

Optimizing mixing time to reduce energy consumption would create a more energy efficient anaerobic digestion (AD) process with higher biogas yield. This study investigates the effect of different mixing strategies on anaerobic digestion of food waste to make the AD waste-to-energy process more energy efficient. Results showed that intermittent mixing is an alternative strategy to continuous mixing or unmixing for high efficient biogas production and energy saving. Through computational fluid dynamics modeling of fluid flow in anaerobic digesters, the mixing time was optimized to 2 mins/hr, at which point the reaction mixture is almost entirely homogeneous. At an organic loading rate of 2.4 g VS_{FW}/L/day, the semi-continuously mixed reactor maintains a higher specific methane yield of 437 ml CH_4/g VS_{FW} in comparison with the other controls. The energy balance investigated the electricity generated and the net heat output generated, in addition to self-sustaining and meeting the energy requirements of the various AD processes investigated. Based on the analysis, it was found the semi-continuous mixing is more energy efficient and sustainable to generate sufficient biogas output for the energy system to provide a net positive heat and electricity output. The original digestate from this 1000L anaerobic digester, heat treated digestate and chemical fertilizer are used for cultivation of vegetables. The results showed that the interaction between heat treatment and digestate concentration is significant, so the trend-lines for the different types of digestate are different. The treatment with the highest yield was also compared with the control treatment (chemical fertilizer, 15N:15P:15K, 1.2 g per application) using t-test. The yield of the best performing digestate is not significantly different from the control. It means that the performance of digestate is almost same to the use of chemical fertilizer.

Topics:
Bioenergy
Environmental Sustainability
Food (Sustainability & Environment)

Preparation of Nanocarbon from Bioenergy Waste Derived Biochar through Mechanized Grinding for the Removal of Pharmaceuticals from Aqueous Media

- Conference: E2S2 CREATE and AIChE Waste Management Conference
- Year: 2019
- Proceeding: 2019 E2S2-CREATE and AIChE Waste Management Conference
- Group: Abstract Submissions
- Session: Section D - Topic 4, Oral Presentation
- Time: Wednesday, March 13, 2019 - 9:15am-9:30am

Authors:
Ramanayaka, S., Ecosphere Resilience Research Center, Faculty of Applied Sciences, University of Sri Jayewardenepura
Jayawardhana, Y., National Institute of Fundamental Studies
Galagedara, H., National Institute of Fundamental Studies
Ok, Y. S., Korea Biochar Research Center, O-Jeong Eco-Resilience Institute (OJERI) & Division of Environmental Science & Ecological Engineering, Korea University
Wijayasinghe, A., National Institute of Fundamental Studies
Vithanage, M., Ecosphere Resilience Research Center, Faculty of Applied Sciences, University of Sri Jayewardenepura

Biochar has received recent attention as an excellent adsorbent in the removal of emerging contaminants from water. Similarly, high surface area of the nanoparticles increase the adsorption capacity of the material. Therefore the objective of this study is to prepare nanocarbon from bioenergy waste derived biochar through mechanized grinding. Dendro biochar, which is a solid waste from the dendro-power industry was used to prepare nanocarbon for the removal of antibiotic Oxytetracycline (OTC) from aqueous media. Biochar was mixed with ethanol and ball milled for 6 hrs around 5000 rpm with 30 minutes intervals for every 1.5 hrs. Resulted nanocarbon was characterized by using powder x-ray diffraction (PXRD), Fourier transform infrared spectroscopy (FTIR) and Nanoscale Particle Size Analyzer. Adsorption edge experiment was carried out from pH 4 to 9 at a dosage of 1 g/L nanocarbon and antibiotic oxytetracycline was analyzed by using UV/Vis spectrophotometer at the wavelength of 354 nm. The particle size of nanocarbon was in the range of 15 to 414 nm. Adsorption of 17 mg/L of oxytetracycline in 24 hrs was observed at pH 4 from a solution with 25 mg/L of concentration. The PXRD pattern for biochar confirmed the amorphous nature of biochar and a poorly crystalline solid, which further revealed a noticeable reflection at 2q = 24° that shows the presence of SiO_2 in the material. Transmittance spectra for the pristine biochar evidenced the functional groups present in the material. Intense peaks at the fingerprint area is accountable due to the presence of aromatics (1650 cm^{-1}) and the presence of C-H bending at 1000 cm^{-1} which confirmed further the aromaticity in the biochar. Hence the results indicate that the nanocarbon produced by dendro biochar is being promising for the removal of OTC from aqueous media.

Topics:
Nanomaterials

Comparison of Different Start-up Strategies for Thermophilic Anaerobic Digestion: Process Stability and Microbial Community Structure Shifts

- Conference: E2S2 CREATE and AIChE Waste Management Conference
- Year: 2019
- Proceeding: 2019 E2S2-CREATE and AIChE Waste Management Conference
- Group: Abstract Submissions
- Session:
  Section D - Topic 4, Oral Presentation
- Time:
  Wednesday, March 13, 2019 - 9:30am-9:45am

Authors:
Lim, J. W., NUS
Tong, Y. W., National University of Singapore

The aim of this study was to compare between step-wise and one-step increase in temperature from mesophilic (35°C) to thermophilic (55°C) for better start-up of thermophilic anaerobic digestion (AD) from a mesophilic digester. The stability of the reactors as well as shifts in microbial community structures were investigated. Stable thermophilic AD (biogas yield of more than 500mL/L/d) was achieved within 10 days from a mesophilic digester treating food waste by adopting the one-step start-up strategy. On the other hand, biogas yield of thermophilic AD with the step-wise start-up strategy dropped to less than 200mL/L/d after the temperature was raised beyond 50°C. After the increase of temperature, thermophilic methanogenic community was established, which was characterised by the colonisation of *Thermotogae*, *Methanosarcina*, *Methanomassiliicoccaceae* and *Methanoculleus*. Results from this study demonstrated that the one-step start-up strategy could allow the rapid establishment of the thermophilic anaerobic microbial community.

Topics:
Bioenergy

Technical and Economic Aspects of Hot Syngas Purification for Power Generation from Gasification of Municipal Solid Waste

- Conference: E2S2 CREATE and AIChE Waste Management Conference
- Year: 2019
- Proceeding: 2019 E2S2-CREATE and AIChE Waste Management Conference
- Group: Abstract Submissions
- Session:
  Section D - Topic 4, Oral Presentation
- Time:
  Wednesday, March 13, 2019 - 9:45am-10:00am

Authors:
Chan, W. P., Nanyang Technological University
Veksha, A., Nanyang Technological University
Dou, X., Nanyang Technological University
Lei, J., Nanyang Technological University
S/O Tamilselvam, K., Nanyang Technological University
Oh, W. D., Universiti Sains Malaysia
Pan, Z. H., Nanyang Technological University
Giannis, A., Nanyang Technological University
Lisak, G., Nanyang Technological University
Lim, T. T., Nanyang Technological University

Gasification is an advanced thermochemical process, which can convert municipal solid waste (MSW) into gaseous fuel (syngas) for utilization in solid oxide fuel cells (SOFCs), gas engines, combined cycle gas turbines (CCGTs) and/or hybrid systems with potentially higher electrical efficiency and lower emission of pollutants. However, this requires the development of cost-efficient syngas purification technologies that are able to remove particulates, tar, HCl, alkali chlorides and sulphur species at high temperatures. The objective of this study was to investigate the technical feasibility and cost benefit analysis of hot syngas treatment using a multi-stage purification process. To achieve these goals, a syngas purification system (5 L/min syngas treatment capacity) was integrated with a lab-scale downdraft gasifier operating on refuse-derived fuel (RDF). RDF pellets were gasified at 850 °C with equivalence air ratio of 0.3 and 25% moisture content to produce syngas with consistent properties. The technical feasibility of the purification process of the raw syngas was evaluated with the focus on the performance of fluidized bed tar reforming (850 °C) and syngas desulfurization (400 °C). Cost effectiveness and energy efficiency of the developed system were analysed using simulated models (ASPEN Plus®). Hot syngas purification is demonstrated to be technically feasible based on the experimental results and economically feasible based on the modeling results.

Topics:
Alternative Energy
Catalysis
Chemical Reaction Engineering

A Data-Driven Two-Stage Resource Allocation Model for Sewage Networks Against Failures

- Conference: E2S2 CREATE and AIChE Waste Management Conference
- Year: 2019
- Proceeding: 2019 E2S2-CREATE and AIChE Waste Management Conference
- Group: Abstract Submissions
- Session:
  Section D - Topic 4, Oral Presentation
- Time:
  Wednesday, March 13, 2019 - 10:00am-10:15am

Authors:
Jiang, J., Shanghai Jiao Tong university
Liu, X., Shanghai Jiao Tong University

Sewage is a type of waste that is produced by domestic households, industrial and agricultural practices. The management of sewage networks is an important environmental issue, because the failure of network components results in sewage overflow to the ground surface, and thus creates offensive odors or even causes health hazards. In order to prevent such component failures, we address the optimal allocation of the limited resources by developing a data-driven two-stage non-linear programming model. In the first stage, the objective is to minimize the total allocated resources and the expected recovery costs of the network and environment, considering the random behaviors of component failures. In the second stage, the amount of leaked sewage caused by component failures is analyzed based on the hydraulic simulation model. The collected data is the historical failure data and detected network state for a fixed inspection interval. Driven by the data, the failure rate of network components is updated by the maximum likelihood estimation. Besides, genetic algorithm is proposed to analyze the two-stage model. In order to demonstrate the practical applicability of the proposed model, the case of a sewage network is conducted and some suggestions on the optimal resource allocations are finally provided. Furthermore, we discuss the impacts on the total costs caused by the data-driven updating of failure rate.

Topics:
Consequence Modeling
Emergency Management
Water (Sustainability & Environment)

Clay-Biochar Nanocomposite As a Media for the Removal of Oxytetracycline Antibiotic from Water

- Conference: E2S2 CREATE and AIChE Waste Management Conference
- Year: 2019
- Proceeding: 2019 E2S2-CREATE and AIChE Waste Management Conference
- Group: Abstract Submissions
- Session:
  Section D - Topic 4, Oral Presentation
- Time:
  Wednesday, March 13, 2019 - 10:15am-10:30am

Authors:
Vithanage, M., Ecosphere Resilience Research Center, Faculty of Applied Sciences, University of Sri Jayewardenepura
Weerasooriyagedara, M., Ecosphere Resilience Research Center, Faculty of Applied Sciences, University of Sri Jayewardenepura
Ashiq, A., Ecosphere Resilience Research Center, Faculty of Applied Sciences, University of Sri Jayewardenepura
Sarkar, B., Department of Animal and Plant Sciences, The University of Sheffield
Ok, Y. S., Korea Biochar Research Center, O-Jeong Eco-Resilience Institute (OJERI) & Division of Environmental Science & Ecological Engineering, Korea University

Oxytetracycline (OTC) is considered as one of the frequently used antibiotic in human and in the veterinary industries, release into environment and recorded in the wastewaters. Biochar has been received recent attention as a material for removing pharmaceuticals in aqueous media however, still the capacity needs to be increased further. This study investigates OTC sorption from water using nanocomposite derived from municipal solid waste biochar (MSW-BC) and montmorillonite. The adsorbents were characterized using powder X-Ray Diffraction (PXRD), Fourier-Transform Infrared (FTIR) spectroscopy, and Scanning Electron Microscope (SEM) before and after adsorbent treatment with OTC. Batch experiments were conducted to study the mechanisms involved in the adsorption with changing pH and time for 10 mg/L initial concentration of OTC at a pH range of 3.5-10 until 10 hrs. The SEM confirmed the successful binding of montmorillonite (MMT) onto the MSW-BC through a flaky structure along the porous morphology. Encapsulation of MMT onto MSW-BC was exhibited through changes in basal spacing of MMT via PXRD analysis. Results from FTIR spectra indicated the presence of functional groups for both pristine materials and composites that are involved in the adsorption. There was no considerable effect from the acidic pH while the basic pHs reduced OTC adsorption. Adsorption capacity ranged in 4-4.5 mg/g. Adsorption reached equilibrium at 8 hrs. The results indicated that the clay-biochar nano composite can be used for the removal of OTC in aqueous media.

Topics:
Nanomaterials

Green Preparation of Fluorescent Carbon Dots from Durian Shell Waste for Highly Sensitive Detection of Fe3+ Ions

- Conference: E2S2 CREATE and AIChE Waste Management Conference
- Year: 2019
- Proceeding: 2019 E2S2-CREATE and AIChE Waste Management Conference
- Group: Abstract Submissions
- Session:
  Section D - Topic 4, Oral Presentation
- Time:
  Wednesday, March 13, 2019 - 10:30am-10:45am

Authors:
Jayaweera, S., Nanyang Technological University
Hu, X., Nanyang Technological Universtiy
Yin, K., Nanjing Forestry University
Ng, W. J., Nanyang Technological University

Carbon dots (CDs) are a novel class of nanomaterials that provide unique optical and electronic properties including tunable photoluminescence, low photobleaching, easy functionalization, aqueous solubility and low toxicity. Production of CDs from cellulose biomass precursors has gained attention over the years due to positive economic and environmental impact. However, the photoluminescence yield of cellulose-based CDs is low in the absence of modification additives. In this study, CDs were produced from Durian shell waste (DSW) in a one-pot process without any modification agents with a mass yield of 13.9% and quantum yield of 6.2%. This was two times higher compared to that of CDs derived from pure cellulose. 90% of the DSW comprised cellulose, hemicellulose and lignin while the remainder consisted of extractives such as carboxylic acids, phenolic compounds, amino acids, aldehydes and esters. These extractives enhanced the oxygenated functional groups in DSW improving the photoluminescence of CDs. The maximum emission was at 436 nm at an excitation wavelength of 360nm and was stable over a wide range of pH and chloride concentration and was resistant to photobleaching. The particles were well dispersed and quasi-spherical with an average particle diameter of 3.2 nm. X-ray photoelectron spectroscopy (XPS) and Fourier transform infrared (FTIR) spectroscopy confirmed the presence of oxygen and nitrogen functional groups on the surface.

The CDs showed high selectivity towards Fe^{3+} and a linear relationship was developed over a concentration range of 0-20 μM with a detection limit of 128 nM (S/N=3). Three actual water samples (lake water, sea water, tap water) were used to validate the model and the relative recoveries of 98.9 -108.6% and an RSD of <4.11%.

Topics:
Nanomaterials
Sensors

Making Use of Wasted Coconut Water through Freeze Concentration

- Conference: E2S2 CREATE and AIChE Waste Management Conference
- Year: 2019
- Proceeding: 2019 E2S2-CREATE and AIChE Waste Management Conference
- Group: Abstract Submissions
- Session:
  Section D - Topic 4, Oral Presentation
- Time:
  Wednesday, March 13, 2019 - 10:45am-11:00am

Authors:
Badilla, D., Philippine Nuclear Research Institute

In its natural form, coconut water spoils easily and causes pollution in rivers and streams where it is being thrown away. Through freeze concentration, the volume is decreased and the coconut water is stabilized making it viable for uses not only as a health drink but also as a short-term intravenous hydration fluid, as a cure for renal disorders, as a laxative and other uses. In this study, coconut water was partially frozen and the ice crystals were separated to produce concentrated coconut water. The objectives of this study are to determine the feasibility of freeze concentration of coconut water and to improve the design of the freeze concentrator. Results showed that large ice crystals produced through slow freezing adsorbed less solute and gave a higher concentration of the coconut water concentrate. Coconut water at lower concentration generally freezes faster and therefore requires a shorter freezing time than those at higher concentrations. Only a difference of 10 Brix or larger gave an appreciable change in the freezing time. The increase in concentration of coconut water through freezing was about 1 Brix for every 200 g of ice crystals removed in a 1 liter sample of partly-frozen coconut water. The desired increase in concentration is largely dependent on the amount of ice crystals removed. Empirical equations were obtained to show the effect of weight of ice crystals on concentration of the coconut water concentrate and of the ice crystals. Freeze concentration is an effective method of concentrating coconut water and produces a useful material. Based on the results of the study, a design of the freeze concentration process and the corresponding equipment was proposed. An improved design has been made after the previous study on the scale up of the initially designed freeze concentrator.

Topics:
Materials

Pinch Analysis and Emission Intensity for Waste Management Planning: EU 28

- Conference: E2S2 CREATE and AIChE Waste Management Conference
- Year: 2019
- Proceeding: 2019 E2S2-CREATE and AIChE Waste Management Conference
- Group: Abstract Submissions
- Session:
  Section D - Topic 4, Oral Presentation
- Time:
  Wednesday, March 13, 2019 - 11:00am-11:15am

Authors:
Fan, Y. V., Brno University of Technology
Klemes, J. J., University of Pannonia

Waste recovery and disposal are one of the biggest management challenges. An adequate waste management design is critical in contributing to the development of a sustainable circular economy. There is a need for a systematic and preferably graphical approach to assessing the emission associated with waste treatment processes followed by strategies for mitigation. This study aims to propose a graphical approach in identifying the appropriate waste management system with lower emissions. The proposed graph-based approach, termed as Extended-Waste Management Pinch Analysis (E-WAMPA) is analogous to the existing WAMPA. It differs from WAMPA by three major points a) Emission intensity of waste management system (Net GHG per capita) is introduced to replace the original y-axis, b) the emission of recycling is not assumed as zero c) The demonstration is based on the defined targets, projection and power grid mix of EU. A generic methodology of E-WAMPA is first presented followed by EU-28 case study to elucidate the application. The considered waste type is municipal solid waste (MSW) and the assessed emission is GHG. E-WAMPA capable to suggest the strategies to fulfill the targeted waste management system intensity of a region (e.g. zero) while meeting the individual treatment targets (e.g. 10 % reduction of landfill target and 65 % recycling rate). One of the possible strategies is by adjusting the waste management system of Malta by having 50 kt of MSW for recycling, 55 kt to incineration, 219 kt to the landfill instead of 21 kt for recycling and 283 kt to landfill. The strength, limitation and way forward of E-WAMPA has also been discussed.

Topics:
Energy (Sustainability & Environment)

Residential Waste Dumping Behaviors for Waste Sorting and Relevant Managerial Insights

- Conference: E2S2 CREATE and AIChE Waste Management Conference
- Year: 2019
- Proceeding: 2019 E2S2-CREATE and AIChE Waste Management Conference
- Group: Abstract Submissions
- Session:
  Section D - Topic 4, Oral Presentation
- Time:
  Wednesday, March 13, 2019 - 11:15am-11:30am

Authors:
Jiang, P., Shanghai Jiao Tong University
Liu, X., Shanghai Jiao Tong University
Cai, Y., Nanyang Boshi Xinju Residential Committee
Jiang, M., Nanyang Boshi Xinju Residential Committee, Jiangchuan Road Street, Minhang District

Waste sorting at sources of individual residential areas is a popular strategy worldwide for improving the sustainability of megacities. However, few effects can be produced if the strategy is implemented by relying solely on propaganda and civic awareness. Along with the strategy, mandatory regulations of 'fixed waste dumping sites and fixed dumping time windows' and encouraging bonuses for 'accumulating points in green accounts' have been implemented at pilot residential communities in Shanghai in 2018. Impressive effects regarding waste sorting and recycling have been achieved in these communities under such measures. In early stages after policy regulations, the general waste dumping behaviors of whole residents are nearly unknown, which poses great challenges for decision-makers to determine appropriate waste dumping time windows and schedule manpower and material resources for daily waste collection operations. Traditional simulation approaches have difficulties for the problem due to the population scale and the complexity of individual behavior. To cope with the problem, we design an analytical framework for residential waste dumping behaviors based on the internet of things and data-mining technologies. In the data-mining component of the framework, waste dumping behaviors are visually illustrated by the probability distributions of dumping time, dumping time interval, dumping amounts and dumping frequency, the time-serial trends of dumping amounts, and the weekly patterns and holiday patterns of dumping amounts. A real case of a pilot residential community in Minhang district of Shanghai is employed to demonstrate the effectiveness of the proposed analytical framework. Based on the study of residential waste dumping behaviors, relevant managerial insights can be effectively offered for decision-makers and practitioners.

Topics:
Environmental Sustainability
Management
Social Aspects (Sustainability & Environment)

Microplastics Bound Transport of Antibiotics in Aquatic Environment

- Conference: E2S2 CREATE and AIChE Waste Management Conference
- Year: 2019
- Proceeding: 2019 E2S2-CREATE and AIChE Waste Management Conference
- Group: Abstract Submissions
- Session:
  Section D - Topic 4, Oral Presentation
- Time:
  Wednesday, March 13, 2019 - 11:30am-11:45am

Authors:
Atugoda, T., Postgraduate Institute of Science, University of Peradeniya, Peradeniya, Sri Lanka
Wijesekara, H., Faculty of Applied Sciences, Sabaragamuwa University of Sri Lanka, Belihuloya, Sri Lanka
Bolan, N., Global Centre for Environmental Remediation (GCER), Advanced Technology Centre, Faculty of Science, The University of Newcastle, Callaghan, NSW, 2308, Australia
Vithanage, M., Ecosphere Resilience Research Center, Faculty of Applied Sciences, University of Sri Jayewardenepura

Microplastics (MP) are considered as potential transport vectors for organic contaminants and heavy metals in the aquatic environment. The emerging contaminants such as pharmaceuticals may sorb onto MP and may influence by different environmental factors and physicochemical properties of MP. The study investigates the sorption behavior of Ciprofloxacin (CPX) antibiotic on polyethylene microplastics (PEMP) through batch sorption experiments. The effects of ionic strength and dissolved organic matter (DOM) were examined by introducing $NaNO_3$ and Humic acid (HA) in to the medium. Pure PEMP was characterized by FTIR to analyze the physicochemical properties of the adsorbent. Batch sorption experiment results showed that the sorption of CPX with pH had a gradual increase reaching a maximum sorption at pH 6-7 and then decreased which is likely due to the speciation of CPX at different pH. In the presence of 0.001 M and 0.1 M $NaNO_3$ ionic strength, the overall CPX sorption capacity of PEMP decreased as the ionic strength increased. Sorption capacity of CPX by PEMP decreased significantly in the range of 0.5 to 2.5 mg L^{-1} of HA highlighting the influence of DOM in the aquatic environment towards the CPX sorption and mobilization by PEMP. Results of the FTIR spectrum attributing to $-CH_2$ and $-CH_3$ functional groups exhibit non-polar hydrophorbic properties of PEMP indicating that hydrophobic interactions between CPX and PEMP have been prominent around neutral pH. Hence, PEMP could be potential vectors to transport CPX in natural aquatic environment where the sorption mechanism is being influenced primarily by the characteristics of the water system and the properties of the adsorbate.

Topics:
Adsorption
Pharmaceuticals (Chemicals & Materials)
Water (Sustainability & Environment)

Three-Stage Anaerobic Co-Digestion of Food Waste and Waste Activated Sludge: Identifying Bacterial and Methanogenic Archaeal Communities and Their Correlations with Performance Parameters

- Conference: E2S2 CREATE and AIChE Waste Management Conference
- Year: 2019
- Proceeding: 2019 E2S2-CREATE and AIChE Waste Management Conference
- Group: Abstract Submissions
- Session:
  Section D - Topic 4, Oral Presentation
- Time:
  Wednesday, March 13, 2019 - 11:45am-12:00pm

Authors:
Zhang, L., National University of Singapore
Loh, K. C., National University of Singapore
Zhang, J., National University of Singapore
Mao, L., National University of Singapore

Food waste (FW) and waste activated sludge (WAS) are critical global issues currently. To combine the benefits of high-solids anaerobic digestion (AD) and wet AD of FW and WAS to improve digester performance and methane production, a three-stage anaerobic digester was developed in this study. Three-stage anaerobic co-digestion of FW and WAS was conducted while corresponding bacterial and methanogen communities were characterized. Results showed that the average methane yield in the three-stage digester was 13-52% higher than that of one- and two-stage digester. An increase of 12-47% in volatile solids reduction was achieved in the three-stage digester, compared to the controls. Bacterial phyla *Proteobacteria, Firmicutes and Bacteroidetes* dominated in one-, two- and three-stage digester while genera *Pseudomonas*, *Tissierella*, and *Petrimonas* were selectively enriched in the three-stage digester due to functional segregation. Taxonomic analysis identified 8 dominant methanogen genera, of which *Methanosarcina*, *Methanosaeta*, *Methanobacterium* and *Methanolinea* collectively accounted for 80%. With increasing organic loading rate and digester stage number, the dominant methanogenic pathway shifted from hydrogenotrophic pattern to acetoclastic pattern and reached a final synergy of these two. *Methanosarcina* was enriched by 1.5-1.7 times in the three-stage digester, contributed to the enhanced methane production. The three-stage digester is promising to be applied to the industrial biogas plants after scale-up and further optimization.

Topics:
Biochemical Processes
Bioenergy
Reactors

Behavior of Recycled Aggregate Reinforced Concrete Columns Under Uniaxial Loading

- Conference: E2S2 CREATE and AIChE Waste Management Conference
- Year: 2019
- Proceeding: 2019 E2S2-CREATE and AIChE Waste Management Conference
- Group: Abstract Submissions
- Session:
  Section D - Topic 4, Oral Presentation
- Time:
  Wednesday, March 13, 2019 - 12:00pm-12:15pm

Authors:
Ali, S. M. S., Exponent
Sajid, H., Exponent
Mustafa, M., Consult-Tech
Khan, A. U. R., NED University of Engineering and Technology
Fareed, S., NED University of Engineering and Technology

The root cause of waste generation is due to rapid expansion in the construction industry, which includes the construction of new buildings and demolition of existing structures. It has drawn the attention towards the limited natural materials for construction specifically aggregates which need to be used wisely to diminish detrimental environmental condition. To counter this problem recycling and reusing of coarse aggregate from waste concrete has been identified as one of the solution. This study focuses on the behavior of reinforced concrete columns made up of recycled aggregate, under uniaxial loading. It determines numerous aspects which may affect the behavior of recycled aggregate concrete; it includes percentage replacement of recycled aggregates (30%, 50%, 70% and 100%) and 28 days curing at a constant w/c of 0.43. Initially, mechanical properties of concrete made up of recycled coarse aggregates were determined along with comparing them with properties of natural coarse aggregates concrete. A slight decrease is observed in all the obtained results. The result shows that with the increase in the percentage of RCA, the load carrying capacity of column also decreases. A pure compression failure was observed in RAC columns. This study provides a basis of using RCA in structural concrete by studying the structural behavior of RC columns and recommends further investigations on the failure mechanism of RC compression and flexural members along with studying its durability characteristics to promote the design of green and sustainable structures leading to environmental sustainability.

Topics:
Design/Green Engineering
Environmental Sustainability

Phytoremediation Plant

- Conference: E2S2 CREATE and AIChE Waste Management Conference
- Year: 2019
- Proceeding: 2019 E2S2-CREATE and AIChE Waste Management Conference
- Group: Abstract Submissions
- Session:
  Section E - Topic 5, Oral Presentation
- Time:
  Wednesday, March 13, 2019 - 1:45pm-2:00pm

Authors:
Mathur, P., HKUST AICHE Student Chapter
Hoque, A., HKUST AICHE Student Chapter
Kandasamy, K., HKUST AICHE Student Chapter
Poyyamozhi, S., HKUST AICHE Student Chapter

Abstract

An increasing world's population has started to put a greater burden on our fresh water resources of which the majority is used to satisfy humanities agricultural needs while increasing amounts of sewage from an exponentially urbanizing society goes unrecycled. Current sewage treatment methods are inefficient when it comes to agricultural purposes as they strip the water of phosphates and nitrates, crucial compounds for plant growth. We propose the use of phytoremediation, the process of chemically purifying water using plants, as the process of choice to purify wastewater. This method removes all chemical compounds while retaining phosphates and nitrates. We have designed an easily scalable low-cost structure that would purify sewage by running it through multiple levels of the water hyacinth plant. A continuous process, the resulting affluent would remove toxins such as heavy metals and reduce BOD and COD to levels suitable for agriculture while retaining phosphates and nitrates optimal for plant growth. This structure requires very little oversight and maintenance. This promises a low-cost method of producing nutrient rich water in rural areas and places suffering from drought while freeing up fresh water resources to satisfy humanity's needs.

Topics:
Environmental Sustainability
Plant Operations
Water (Sustainability & Environment)

Life-Cycle Assessment and Optimization Framework for Sustainable Urban Farming Systems

- Conference: E2S2 CREATE and AIChE Waste Management Conference
- Year: 2019
- Proceeding: 2019 E2S2-CREATE and AIChE Waste Management Conference
- Group: Abstract Submissions
- Session: Section E - Topic 5, Oral Presentation
- Time: Wednesday, March 13, 2019 - 2:00pm-2:15pm

Authors:
Li, L., National University of Singapore
Wang, X., National University of Singapore
Li, X., National University of Singapore
Wang, C. H., National University of Singapore
Chong, C., Technoponics Pte Ltd

With the growing population and continuous urbanization, sustainable and secure food production is increasingly important in future urban development. Various advanced farming systems have been developed. The economic performance and environmental impact are important decision criteria for the deployment of such systems. In order to assess and improve the sustainability of urban farming systems, a holistic decision-support framework combining life cycle assessment and optimization is proposed in this work.

After defining the goal and scope of the life cycle assessment and optimization, the life cycle inventory that contains the data of energy and material flows around the system boundary in different phases of the system life cycle is tabulated as the data input for the assessment and optimization of the farming system. The economic performance and environmental impact are evaluated and modeled using the life cycle assessment approach and serve as two objective functions in the optimization of the farming system. Moreover, constraints in the optimization are required to be formulated based on the local market conditions (e.g. demand and price of products, the price of equipment and utilities, and land availability). Decision variables in the optimization including the design of the farming facility, types and amount of crops to be planted and the operating conditions of the farming system can be determined to maximize the economic benefit and minimize the negative environmental impact.

A case study on the design and assessment of a novel modularized urban leafy-vegetable farming system in Singapore utilizing fertilizer derived from beer residue is carried out as a demonstration of the proposed framework. The result of the case study shows the optimized modular vertical farm is promising to be adopted in the urban area for multiple benefits including green and economical urban vegetable supply, waste valorization, and efficient land utilization.

Topics:
Case Studies (Sustainability & Environment)
Computing and Systems Engineering
Food (Sustainability & Environment)

Market Potential of Biomethane in South Korea

- Conference: E2S2 CREATE and AIChE Waste Management Conference
- Year: 2019
- Proceeding: 2019 E2S2-CREATE and AIChE Waste Management Conference
- Group: Abstract Submissions
- Session:
  Section E - Topic 5, Oral Presentation
- Time:
  Wednesday, March 13, 2019 - 2:15pm-2:30pm

Authors:
Kim, W., Kyungpook National University
Jo, J. H., Korea Environment Institute

This study estimated domestically available energy amount of biomethane including landfill gas (LFG) as a transportation fuel by 2035. The amount of available energy that could be supplied was predicted through four stages of 'theoretical-', 'geographical-', 'technical-' and 'market potential' by considering geographical, technical, economic conditions, etc. Energy efficiency and added value of biomethane are largely influenced by the site conditions and the neighboring infrastructures. So, how much of the natural gas used in transportation could be substituted with biomethane was examined by setting limits to the amount of organic wastes generated within urban areas. As a result, the market potential of biomethane including landfill gas was approximately 331 x 10^6 Nm^3/yr, corresponding to 25% of the natural gas supply for transportation, which could be replaced by biomethane. Assuming that 2% of natural gas for transportation is replaced by biomethane, it corresponds to 29 x 10^6 Nm^3/yr (approximately 9% of market potential of biomethane). However, RFS annual mixing rate may be increased upon introduction and the growth rate of the natural gas supply for transport would be higher than that of market potential of biomethane calculated in this study. Therefore, it is vital to ensure a stable supply of biomethane, to improve efficiency in technical development and management, and to lay out specific quality requirements for promoting the use of biomethane as a transportation fuel for efficient implementation of RFS policy.

Topics:
Bioenergy

Preparation of a Novel Composite Using Chitosan, Laterite and Iron Oxide Nanoparticles and Its Efficacy in Removing Methylene Blue Dye and Arsenic from Water

- Conference: E2S2 CREATE and AIChE Waste Management Conference
- Year: 2019
- Proceeding: 2019 E2S2-CREATE and AIChE Waste Management Conference
- Group: Abstract Submissions
- Session:
  Section E - Topic 5, Oral Presentation
- Time:
  Wednesday, March 13, 2019 - 2:30pm-2:45pm

Authors:
Sakib, M. N., University of Dhaka
Ahmed, S., Bangladesh University of Engineering and Technology

Bio-materials are abundant and are finding new avenues of applications. In this research, chitosan, iron oxide nano-particles and laterite clay composite was prepared using sodium-polyphosphate as the cross linker. The reasons of using chitosan are that it is easily available in our country, often used as adsorbents and is biodegradable. Iron oxide nano particles were used to recover the composite. Furthermore, they are often used for the adsorption of arsenic. Laterites are common in Bangladesh and is cheap. Hence, the composite compositions were chosen with a view to preparing them in large scale cost effectively. Composites with different compositions of the three ingredients were prepared and their pore volume and methylene blue (MB) adsorption capacity was compared and the study shows that 1:1:1 composition provides the best suitable adsorption. Removal of MB using this composite shows that with the increase of pH the removal increases and showed maximum adsorption capacity (Q) of 13 mg/g and for arsenic the pH dependency follows the opposite trend of MB and Q value is 0.291 mg/g. Both cases shows that the adsorption is following Langmuir Isotherm model and chemi-sorption is the rate determining step. Studying the nature of the composite tells us that it can be a good anti-microbial agent as well. Finally this non-toxic composite harbors potential in the fields of application of different branches of environmental and pharmaceutical science.

Topics:
Adsorption
Biomaterials
Water (Sustainability & Environment)

Batch and Fixed Bed Column Based Optimization of Municipal Solid Waste Biochar Application for Volatile Organic Compound Removal Form Landfill Leachate

- Conference: E2S2 CREATE and AIChE Waste Management Conference
- Year: 2019
- Proceeding: 2019 E2S2-CREATE and AIChE Waste Management Conference
- Group: Abstract Submissions
- Session:
  Section E - Topic 5, Oral Presentation
- Time:
  Wednesday, March 13, 2019 - 2:45pm-3:00pm

Authors:
Jayawardhana, Y., National Institute of Fundamental Studies
Vithanage, M., Faculty of Applied Sciences, University of Sri Jayewardenepura

Municipal solid waste biochar (MSW-BC) for the removal of volatile organic compounds (VOCs) from aqueous media was investigated in this study. Batch result supported fixed bed column study was conducted for optimization of site specific permeable reactive barrier. The characteristics of MSW-BC were evaluated under different categories, surface properties, proximates, and elemental. Batch studies for BTEX conducted in different ranger of sorbent dosages (1-10 g/L), pH conditions (3, 5, 7, 9), time intervals (0.5- 24 h) and final sorbate concentrations (10- 600 µg/L). The batch based column experiments conducted for VOCs like toluene and m-xylene. Different quantity from MSW-BC (1, 0.25 %) was utilized into the column and different inlet concentration of Toluene (1000, 4000 µg/L) Xylene (4000 µg/L) was tested for optimization. According to the MSW-BC characteristics, organic sorption favor due to high carbonized structure of sorbent increasing hydrophobicity depicted by proximate analysis of lower polarity index [(O+N)/C] (0.26). Elemental characteristics of lower bioavailable metal environment ideal for field application of MSW-BC. Maximum adsorption of the Isotherm study reveals higher capacity for all VOCs (400 – 700 µg/g). The model fittings of all isotherms well suited with Langmuir and further Fruendlich fitting describe both chemical and physical adsorption. In fixed bed column system, relatively low r^2 values ($r^2 = 0.8833, 0.8808$) indicated the less applicability of Adams-Bohart model. Hence, both sorbates obey the Thomas model and could explain the Langmuir kinetics adsorption by MSW-BC in a fixed bed system. Hence, it can be concluded that MSW-BC is efficient in the removal of VOCs in batch as well as fixed bed sorption systems simultaneously minimizing the waste acumination in the environment.

Topics:
Adsorption
Water (Sustainability & Environment)

The Application of Machine Learning for Modeling the Adsorption of Heavy Metals on Biochars with Different Origins

- Conference: E2S2 CREATE and AIChE Waste Management Conference
- Year: 2019
- Proceeding: 2019 E2S2-CREATE and AIChE Waste Management Conference
- Group: Abstract Submissions
- Session:
  Section E - Topic 5, Oral Presentation
- Time:
  Wednesday, March 13, 2019 - 3:00pm-3:15pm

Authors:
Zhu, X., National University of Singapore
Wang, X., National University of Singapore
Ok, Y. S., Korea Biochar Research Center, O-Jeong Eco-Resilience Institute (OJERI) & Division of Environmental Science & Ecological Engineering, Korea University

Biochar, as the solid by-product of biomass, had been applied in multidisciplinary area such as wastewater treatment due to its microporous structure and abundant surface functional groups. The treatment efficiency of biochar for wastewater was significantly influenced by biochar characteristics and the environmental conditions. However, the diversity of biomass feedstock and uncertainty of produced biochar made the relationship complicated. Machine learning may be preferred to resolve the problem through teaching machine to find, recognize and extract relationships or rules based on statistical data. The adsorption of six heavy metals (lead, cadmium, nickel, arsenic, copper and zinc) on the fifty kinds of biochars were modeled using artificial neural network and random forest in the study, respectively, based on collected 433 datasets of adsorption experiments from the literatures. The regression models were trained and optimized to predict the adsorption efficiency according to biochar characteristics, metal sources, environmental conditions (e.g. temperature and pH in water and wastewater), and the initial concentration ratio of metals to biochars. The RF model showed better accuracy and predictive performance for adsorption efficiency with lower root mean squared error (0.038) and higher regression coefficient (R^2=0.986) than ANN. The initial concentration ratio of metals to biochars was demonstrated as the most significant factor for adsorption efficiency, while the biochar properties was secondary. The CEC and pH_{H2O} of biochars accounted for over half of the contribution of biochar characteristics and the surface area was only accounted for 2.4% of adsorption. The accurate predicted ability of developed model could decrease experiment workload such as predicting the adsorption efficiency according to the biochar characteristics, environmental conditions, and the target metals. Moreover, the relative importance of each variable could provide a right direction for better removing heavy metals in the real water and wastewater.

Topics:
Adsorption
Biomaterials
Metals

Environmental Impacts from Energy Use in Solar Photovoltaics Manufacturing

- Conference: E2S2 CREATE and AIChE Waste Management Conference
- Year: 2019
- Proceeding: 2019 E2S2-CREATE and AIChE Waste Management Conference
- Group: Abstract Submissions
- Session:
  Section E - Topic 5, Oral Presentation
- Time:
  Wednesday, March 13, 2019 - 3:15pm-3:30pm

Authors:
Quek, A., NUS
Ee, A., NUS

Solar energy have been considered as a clean, renewable and non-pollutive source of energy. However, the environmental impacts of solar energy are usually not considered and not well understood. Crystalline solar photovoltaic (PV) panels are the most popular form of solar energy in use currently. A cradle to gate life cycle assessment (LCA) was done for mono and polycrystalline solar photovoltaic (PV) panels. The raw materials, electricity and other inputs were considered in nine impact categories, up to the manufacturing stage. Results showed that energy input had the greatest environmental impacts among all inputs. Of all possible sources of energy input, coal use had the highest environmental impact. Increasing the proportion of renewable energy use in the manufacturing of PVs reduced the impacts across all categories. This implies that as more PVs are manufactured for use in the coming years, more carbon emissions will be expected, generating a carbon debt, unless a corresponding increase in renewable energy is use in manufacturing. This is also true for other environmental impacts, such as acidification, aquatic, terrestrial and human toxicities. However, certain processes in PV manufacturing, such as silicon ore refining has no known renewable energy substitute. The limits on the use of renewable energy in solar PVs manufacturing would limit the environmental benefits of using PVs and negatively affect decarbonization efforts.

Topics:
Solar Energy

King Coconut Husk Derived Biochar through Hydrothermal Activation Process for the Removal of Cadmium (II) and Copper (II) from Aqueous Media

- Conference: E2S2 CREATE and AIChE Waste Management Conference
- Year: 2019
- Proceeding: 2019 E2S2-CREATE and AIChE Waste Management Conference
- Group: Abstract Submissions
- Session:
  Section E - Topic 5, Oral Presentation
- Time:
  Wednesday, March 13, 2019 - 3:30pm-3:45pm

Authors:
Galagedara, H., National Institute of Fundamental Studies
Vithanage, M., Ecosphere Resilience Research Center, Faculty of Applied Sciences, University of Sri Jayewardenepura
Wijayasinghe, A., National Institute of Fundamental Studies

This research was conducted to study the removal of the most prevalent heavy metal ions cadmium (II) and copper (II) by adsorption on King coconut (Red Dwarf coconut, *cocos nucifera.* var) husk biochar. Raw king coconut biochar (RKH) and steam activated King coconut biochar (AKHC) were prepared and analysed for physical properties, surface properties, elemental composition and functional properties to identify the sorbent suitability. The efficiency of adsorption of heavy metal ions [cadmium (II) and copper (II)] on biochar were studied with batch sorption experiments. Adsorption capacities of RKH and AKHC for cadmium (II) and copper (II) at different pH, initial metal concentrations, ionic strengths and contact times between adsorbent and adsorbate were measured. Scanning electron microscopy reveals the existence of a better and highly porous structure in AKHC than that of RKH, that could be a result of the volatilization by steam process. EDX analysis indicates a carbonized structure for both adsorbents. Further, it was observed that the adsorption of Cd and Cu ions by AKHC could be increased with increasing pH and decreased with ionic strength. Supported point of zero charge pH_{pzc} (5.8, 6.9)can be used to explain the effect of pH on the higher adsorption process at higher pH. The maximum adsorption capacities of Cd and Cu of AKHC were 22.24, 34.24 mg g^{-1}, respectively, and they were considerably lower for RKH with 17.22, 21.25 mg g^{-1}. With well-fitting of Freundlich isotherm, it revealed the heterogeneous properties of the surfaces and favourable adsorption process. The kinetics fittings of pseudo second order were accommodated towards chemisorption process for both Cu and Cd. The higher sorption ability expressed by AKHC and RKH, prepared in this study, show their potential sorbent activity for environmental remediation application.

Topics:
Adsorption

Application of Liquid Phase Plasma in Photocatalytic Hydrogen Production from Wastewaters Containing Organic Pollutants

- Conference: E2S2 CREATE and AIChE Waste Management Conference
- Year: 2019
- Proceeding: 2019 E2S2-CREATE and AIChE Waste Management Conference
- Group: Poster Submissions
- Session:
  Poster Session
- Time:
  Tuesday, March 12, 2019 - 3:55pm-4:20pm

Authors:
Park, Y. K., University of seoul
Jung, S. C., Sunchon National University
Chung, K. H., Sunchon National University

The hazardous wastes and toxic water pollutants have brought out serious environmental problems associated with attracted much attention. In the various physical, chemical and biological techniques for treatment of wastewaters, heterogeneous photocatalysis has been considered as an effective alternative for water remediation. Photocatalytic water splitting is effective for converting water to hydrogen as a clean and renewable energy. It has been also applied on deconposition of organic materias in wastewater. The development of high photosensitive catalysts has been studied under UV and visible light illumination. Although a range of light sources have been employed in photocatalysis, few studies have examined photocatalysis for hydrogen generation using liquid phase plasma by irradiation into water directly.

Photocatalytic decomposion using liquid phase plasma was assesd to produce hydrogen and to remove polluant from wastewater containing organic materials. Characteristics of liquid phase plasma irradiation were examined in the photocatalytic water splitting for hydrogen evolution on modified TiO_2 photocatalysts. Optical emission spectra were characterized with variation of plasma discharging parameters of the liquid phase plasma. Pathway of photocatalytic reaction for hydrogen evolution was also evaluated with degradation of organic pollutants.

The hydrogen was generated in the photodecomposition of water through the liquid phase plasma irradiation without photocatalysts. The active species generated by the liquid phase plasma irradiation brought out the hydrogen evolution regardless of the photocatalysts. The rate of hydrogen evolution depended on the parameters of liquid phase plasma discharging. The hydrogen evolution was significantly enhanced by irradiation of liquid phase plasma from the wastewater. The organic pollutant assisted to improve the hydrogen production with adding of hydrogen by its decomposition.

Topics:
Hydrogen Production & Storage

Case Study - Demonstration Plant of Water Recycling of Treated Industrial Wastewater Using Microfiltration/Reverse Osmosis and Activated Carbon Filter for Process Use in a Petrochemical Plant, Singapore

- Conference: E2S2 CREATE and AIChE Waste Management Conference
- Year: 2019
- Proceeding: 2019 E2S2-CREATE and AIChE Waste Management Conference
- Group: Poster Submissions
- Session:
  Poster Session
- Time:
  Tuesday, March 12, 2019 - 3:55pm-4:20pm

Authors:
Francis, T., Petrochemical Corporation of Singapore (Private) Limited
Ee, B. S., Petrochemical Corporation of Singapore (Private) Limited
Sim, K. S., Petrochemical Corporation of Singapore (Private) Limited

Case study – Demonstration plant of water recycling of treated industrial wastewater using Microfiltration/Reverse Osmosis and Activated Carbon Filter for process use in a petrochemical plant, Singapore

Membrane processes employing pre-treatment with micro-filtration (MF) with reverse osmosis (RO) have been proven to be reliable and economical in Industrial Water Recycling facilities. This has resulted in the conversion of existing Conventional Activated Sludge (CAS) process for treated industrial wastewater recovery for process use.

The aim of this paper is to demonstrate a phased approach of carrying out technological and economic assessment of an alternative approach of applying membrane technology, for water recycling of treated industrial waste water for process use. It has been installed at downstream of the existing CAS at a wastewater treatment plant in the Singapore Petrochemical Complex.

Due to the variability of industrial effluents, this demonstration plant was developed following a rigorous Phase 1 – Pilot Plant with different configuration of pre-treatment membrane (MF/UF) and material selection with RO. Actual treated wastewater effluent was used for the wide range of investigation, particularly on the susceptibility of membranes fouling and formulation of cleaning solutions for membrane flux recovery. The Phase 2 demonstration plant was then designed and constructed, followed by trial runs with enhanced cleaning regimes for fouling control, to validate critical parameters such as allowable flux for a pressurized pre-treatment membrane units and RO permeate quality. The permeate is then passed through an activated carbon filter, to remove any residual organics and free chlorine as the feed water to ion-exchange pure water process unit to produce boiler feed water (BFW) to generate high pressure steam.

Indeed, with the expected successful implementation of the wastewater recovery system, PCS's uptake of fresh water can be potentially reduced by about 10% when the recycling plant operates at its full capacity.

Topics:
Sustainability & Environment

Developing Methodology to Determine the Efficacy of Odor Reduction Technology of Liquid Fertilizer Circular Operation Technique System

- Conference: E2S2 CREATE and AIChE Waste Management Conference
- Year: 2019
- Proceeding: 2019 E2S2-CREATE and AIChE Waste Management Conference
- Group: Poster Submissions
- Session:
  Poster Session
- Time:
  Tuesday, March 12, 2019 - 3:55pm-4:20pm

Authors:
Kim, M. S., Ojeong
Hwang, J. H., Jeonju University
Lee, S. R., Jeonju University

NH_3 emissions from liquid waste were commonly estimated based on thin-layer models. However, such models differ in their ability to accurately reproduce observed emissions, which may be partly due to an incomplete mechanistic understanding of NH_3 volatilization. In this study, NH_3 release from pure solutions of $(NH_4)_2SO_4$ (AS), swine pit liquid (PL), and liquid fertilizer sample (LFS) from the liquid fertilizer operation technique system were evaluated under controlled conditions (pH 7.6 and temperature 20 °C). Relationships between the NH_3 release and pH, temperature, and total ammoniacal nitrogen (TAN) were established. Under the controlled TAN conditions, the estimation of NH_3 emission were used to determine the efficacy of odour reduction when LFS solution is flushed into the pit in swine animal feeding operations with liquid fertilizer circular operation technique system.

Topics:
Ammonia
Mixing
Water (Sustainability & Environment)

Developing Reduction Technology of Ammonium-Salt Based Inorganic Second Aerosol from Livestock Manure Composting Process

- Conference: E2S2 CREATE and AIChE Waste Management Conference
- Year: 2019
- Proceeding: 2019 E2S2-CREATE and AIChE Waste Management Conference
- Group: Poster Submissions
- Session:
  Poster Session
- Time:
  Tuesday, March 12, 2019 - 3:55pm-4:20pm

Authors:
Lee, S. R., Jeonju University
Kim, M. S., Ojeong
Hwang, J. H., Jeonju University
Ok, Y. S., Korea Biochar Research Center, O-Jeong Eco-Resilience Institute (OJERI) & Division of Environmental Science & Ecological Engineering, Korea University

Disturbance to the global nitrogen (N) cycle receives less attention compared with the carbon cycle, which imbalanced more N pollution. The N cycle related with animals and various industrial activities influenced in a complicate manner based on reactive forms of Ammonia (NH_3) and NOx. Atmospheric transfer of reactive nitrogen compounds, especially NH_3 and the relations with balance of fine particulate matter (PM) are solely focused on the domestic effect from the agriculture in South Korea while the external impact on particulate matter is mainly affected from China. Additionally, the methodology of dry/wet deposition network was engaged in our research site with help of US National Atmospheric Deposition Program (NADP) to develop a comprehensive multi-pollutant approach for N, with special attention to the fate of emissions from agriculture. The reduction technology of ammonium-salt based inorganic second aerosol from livestock manure composting process was applied using the cattle manure-based biochar as the compost ingredients of moisture control mixing material.

Topics:
Ammonia
Particle Characterization
Particle Technology

Development of MoS2/Graphene on Carbon Cloth Electrode for Simultaneous Wastewater Treatment and Hydrogen Production Using Water Splitting

- Conference: E2S2 CREATE and AIChE Waste Management Conference
- Year: 2019
- Proceeding: 2019 E2S2-CREATE and AIChE Waste Management Conference
- Group: Poster Submissions
- Session:
  Poster Session
- Time:
  Tuesday, March 12, 2019 - 3:55pm-4:20pm

Authors:
Jeon, K. J., Inha University
Park, Y. K., University of Seoul

In recent year, there has been increasing interest in finding solution for the efficient removal of organic contaminants in wastewater and hydrogen production as an alternative energy. In this study, we developed an electrode which is capable of treating organic compounds in wastewater and generating hydrogen by water splitting at the same time. Carbon cloth (CC) is used as an electrode substrate to overcome cost issue of noble metal electrode substrate and catalysts. To overcome limitations of CC such as low conductivity and low active site, the high electrical conductivity of graphene and the large amount of active sites of MoS_2 were synthesized on CC substrate via chemical vapor deposition method. As the result, MoS_2/graphene-CC electrode showed better hydrogen evolution performance than bare-CC and MoS_2-CC. Moreover, we confirmed that the organic compound in artificial wastewater electrolyte was decomposed after electrochemical reactions. We expected to use MoS_2/graphene-CC electrode for wastewater treatment and hydrogen evolution by electrolysis.

Topics:
Energy (Sustainability & Environment)
Hydrogen Production & Storage
Water (Sustainability & Environment)

Efficient Mg-Doped Porous Carbon Derived from Petroleum Coke for Removing Heavy Metal Ions from Wastewater

- Conference: E2S2 CREATE and AIChE Waste Management Conference
- Year: 2019
- Proceeding: 2019 E2S2-CREATE and AIChE Waste Management Conference
- Group: Poster Submissions
- Session:
  Poster Session
- Time:
  Tuesday, March 12, 2019 - 3:55pm-4:20pm

Authors:
Kim, M. J., Korea University
Lee, K. B., Korea University
Yuan, X., Korea University

Water pollution caused by addition of heavy metal ions resulting from the industrial activities is increasing tremendously and becomes one of the top hottest environmental issues in the entire world. This research focuses on developing an effective adsorption process for treatment of heavy metals-contaminated wastewater. Mg-doped porous carbon was prepared by impregnating Mg onto KOH-activated petroleum coke at a mass ratio of 2, after conducting the KOH activation at 450 °C. The removal of Cd(II), Pb(II), Cu(II), and Ni(II) by the prepared Mg-doped porous carbon was studied depending on initial concentration, pH, contact time, adsorbent dosage and temperature dependent. The removal performance and the selectivity sequence of four mixed heavy metal ions in aqueous solution were also investigated for applying to real industrial wastewater. The adsorption parameters for removing different heavy metal ions were determined using both isotherm and kinetic models. A mechanism for treatment of four mixed heavy metal ions was also investigated. The Mg-doped porous carbon derived from petroleum coke is cost-effective and efficient, and therefore it can be an alternative, practical and promising material for the treatment of wastewater.

Topics:
Adsorption
Nanomaterials
Water (Sustainability & Environment)

High-Throughput Profiling of Antibiotic Resistant Genes in a Novel Tridimensional Eco-Biological WWTP with High Discharge Standards for Reuse

- Conference: E2S2 CREATE and AIChE Waste Management Conference
- Year: 2019
- Proceeding: 2019 E2S2-CREATE and AIChE Waste Management Conference
- Group: Poster Submissions
- Session:
  Poster Session
- Time:
  Tuesday, March 12, 2019 - 3:55pm-4:20pm

Authors:
Yu, K., Shanghai Jiao Tong University
Chen, Y., Shanghai Jiao Tong University
Zhang, B., Shanghai Jiao Tong University
Gin, K. Y. H., National University of Singapore
He, Y., Shanghai Jiao Tong University
Li, P., Shanghai Jiao Tong University
Huang, Y., Shanghai Jiao Tong University

Antibiotic resistome is a raising concern around the world, especially considering reclaimed wastewater. In this study, a novel tridimensional eco-biological wastewater treatment plant (TEB-WWTP) with high discharge standard (CODcr < 20 mg/L, ammonia < 1 mg/L, TP < 0.2 mg/L, third sort of *Environmental quality standards for surface water in China*) for reuse was selected. The bioreactors were characterized by adding internal bio-modules and planting various macrophytes which were fixed by stainless steel cages with additional bio-ceramsites on the top of bioreactors to purify odor gas. WWTPs are considered as hotspots of antibiotic resistant genes (ARGs). High-throughput qPCR (HT-qPCR) was applied here to profile ARGs in the TEB-WWTP and receiving stream. A total of 228 ARG subtypes were detected in the TEB-WWTP system which exhibited good performance for the removal of ARGs with the number, relative abundance and absolute abundance of ARGs largely decreased in the final effluent. Surprisingly, we found that the number and relative abundance of ARGs in the rhizosphere of three macrophytes (*Canna indica, Cyperus alternifolius* and *Musa basjoo*) were lower than that of biological reactor effluent. Aquatic plants probably play an important role in the inhibition of ARGs dissemination. Most of ARGs have significant correlations ($p < 0.001$) with total abundance of mobile genetic elements (MGEs), suggesting that horizontal gene transfer was the main mechanism for ARGs propagation. More attention should be paid to excess sludge with high relative abundance of ARGs for further dispose or farmland application. The final effluent was discharged to a stream for ecological reuse. The number and relative abundance of ARGs in the sediment of downstream were much higher than that of upstream, indicating that the risk of ARGs can't be overlooked when considering wastewater reuse.

Topics:
Water (Sustainability & Environment)

Investigation of Phosphorus Removal in Sewage Sludge Solid Fuel Using Citric and Formic Acid

- Conference: E2S2 CREATE and AIChE Waste Management Conference
- Year: 2019
- Proceeding: 2019 E2S2-CREATE and AIChE Waste Management Conference
- Group: Poster Submissions
- Session:
  Poster Session
- Time:
  Tuesday, March 12, 2019 - 3:55pm-4:20pm

Authors:
Xu, L. H., Ajou university

The amount of sewage sludge continues to increase with expansion of the sewage treatment plant. However, ocean dumping of sewage sludge has been prohibited with the London Protocol for the Prevention of Marine Pollution. Therefore, it is an important issue to treatment of sewage sludge on the ground site. The hydrothermal carbonization technology is one of the attractive solutions to convert the sewage sludge into solid fuel. It can decrease sewage sludge and can produce alternative energy.

Unlike coal, the sewage sludge solid fuel contains large quantities of phosphorus. The phosphorus is commonly known as a substance that significantly impacts fouling and slagging. Fouling and slagging is a factor that lowers energy efficiency. Therefore, it is important to remove phosphorus for reducing fouling and increasing energy efficiency. In this experiment, the sewage sludge solid fuel, which was produced from Incheon city, S.Korea, was used. Citric acid and formic acid was used to dissolve phosphorus in water. Sewage sludge solid fuel mixed in citric and formic solution and stirred for 24hours. In order to investigate phosphorus removal by citric and formic acid, phosphorus-dissolved solution was analyzed by Ion Chromatography. As a result, phosphorus removal efficiency of citric acid is higher than formic acid. In order to investigate fouling property of sewage sludge solid fuel with phosphorus removed, muffle furnace experiment and agglomeration test was conducted.

Topics:
Energy
Environmental Sustainability
Fuels

Mesophilic Versus Thermophilic Anaerobic Digestion of Soybean Curd Residue for Methane Production: Characterizing Bacterial and Methanogen Communities and Their Relationship with Organic Loading Rates

- Conference: E2S2 CREATE and AIChE Waste Management Conference
- Year: 2019
- Proceeding: 2019 E2S2-CREATE and AIChE Waste Management Conference
- Group: Poster Submissions
- Session:
  Poster Session
- Time:
  Tuesday, March 12, 2019 - 3:55pm-4:20pm

Authors:
Zhang, L., National University of Singapore
Loh, K. C., National University of Singapore
Sarvanantharajah, S., University College London

How to economically and effectively treat huge amount of soybean curd residue (SCR) is a current major challenge to the soybean products processing plants throughout the world due to the fact that the uncontrolled disposal of SCR is liable to cause adverse environmental impacts and threaten the public health. To find the optimal operation parameters for anaerobic digestion of SCR, mesophilic (35^0C) versus thermophilic (55^0C) anaerobic digestion (AD) of SCR for methane production were contrastively investigated for 95 days with an increasing organic loading rate (OLR) from 0.55 to 4.96 gVS/L. Anaerobic digestion performance of SCR indicated that the maximum available OLR was 3.3 gVS/L for both mesophilic and thermophilic AD. Thermophilic AD provided a higher methane yield and volatile solids removal than that of mesophilic AD. The first dominant volatile fatty acid was identified as acetic acid. In addition, an in-depth understanding of the microbial compositions, diversity/similarity, and relations between digester performance, biodiversity and environmental parameters at the genome level was obtained by pyrosequencing technique and bioinformatics analysis, which would help to optimize microbial productivity and maximize the methane yield in the anaerobic digestion process of the SCR.

Topics:
Biochemical Processes
Bioenergy
Food (Biological Engineering)

Preparation of Waste Coffee Ground-Based Microporous Carbons for CO2 Capture

- Conference: E2S2 CREATE and AIChE Waste Management Conference
- Year: 2019
- Proceeding: 2019 E2S2-CREATE and AIChE Waste Management Conference
- Group: Poster Submissions
- Session:
  Poster Session
- Time:
  Tuesday, March 12, 2019 - 3:55pm-4:20pm

Authors:
Kim, M. J., Korea University
Lee, K. B., Korea University
Choi, S. W., Korea university

As the amount of CO_2 in the atmosphere increases, global warming deepens and climate change problems are occurring. Therefore, CO_2 capture and storage (CCS) technology is considered one of the ways to reduce the amount of emitted CO_2. In particular, more than 80% of the total cost of CCS technology is accounted by the capturing stage, therefore efficient CO_2 capturing technology is essential. Absorption, adsorption, and membrane separation are mainly used in post-combustion capture technology. Among them, the adsorption has advantages of low energy consumption and easy regeneration. However, its performance is lower than that of the conventional absorption technology. Thus, it is required to develop efficient adsorbents. Among various materials used as an adsorbent, activated carbon is advantageous in that it is easy to synthesize, cost competitive, and easily controlled in pore size. In this study, waste coffee grounds were used as a carbon precursor and porous carbon materials were synthesized by using K_2CO_3 as an activating agent. Based on the discarded coffee grounds in daily life, high value-added microporous carbon was produced by the simple synthesis method. The changes of surface area and CO_2 adsorption performance were analyzed with the activation temperature and time and the correlations between the variables were investigated.

Topics:
Adsorption
Carbon Capture & Storage

Robust Trading Strategies for a Waste-to-Energy Combined Heat and Power Plant in a Day-Ahead Electricity Market

- Conference: E2S2 CREATE and AIChE Waste Management Conference
- Year: 2019
- Proceeding: 2019 E2S2-CREATE and AIChE Waste Management Conference
- Group: Poster Submissions
- Session:
  Poster Session
- Time:
  Tuesday, March 12, 2019 - 3:55pm-4:20pm

Authors:
Hu, C., University of Technology, Sydney
Liu, X., Shanghai Jiao Tong University
Lu, J., University of Technology, Sydney

Waste-to-energy (WtE) technology has been used all over the world as it can solve the dilemma of waste management, energy demand, and global warming. Many modern WtE plants are built and operated in a combined heat and power (CHP) mode due to the high overall energy efficiency. This paper studies robust trading strategies for a WtE CHP plant which sells electricity in a day-ahead electricity market and exports heat to a district heating network. Owing to the requirements of the day-ahead electricity market, plant operators must determine the trading strategy one day before real delivery of electricity. However, many key problem parameters including electricity prices, heat demand, and the amount of waste delivered to the plant are uncertain at the day-ahead stage. To derive robust electricity trading strategies for the WtE CHP plant under different types of uncertainty, a two-stage robust optimization model is developed and a solution procedure based on the column-and-constraint generation method is designed. A case study is also performed to illustrate the effectiveness of the robust model and the solution procedure.

Topics:
Energy Production
Plant Operations

Thermochemical Conversion of Polyvinyl Chloride (PVC) in the Presence of Fe3O4 and CO2 for Syngas Production

- Conference: E2S2 CREATE and AIChE Waste Management Conference
- Year: 2019
- Proceeding: 2019 E2S2-CREATE and AIChE Waste Management Conference
- Group: Poster Submissions
- Session:
  Poster Session
- Time:
  Tuesday, March 12, 2019 - 3:55pm-4:20pm

Authors:
Song, H., Sejong University
Kwon, G., Sejong University
Cho, D. W., Korea Institute of Geoscience and Mineral Resources

Thermochemical energy recovery (*i.e.*, syngas production) from waste plastic valorization is more challenging compared to the recovery from biomass due to the relatively high stability of plastic. Thermal degradation of polyvinyl chloride (PVC) assisted by Fe_3O_4 was carried out to investigate the feasibility of using Fe_3O_4 to facilitate enhanced syngas production (H_2 and CO). The enhanced generation of H_2 and CO from thermolysis of PVC/Fe_3O_4 was more obvious in CO_2 atmosphere than N_2 atmosphere. In the presence of Fe_3O_4, the generation of CO from PVC thermolysis began at 560 °C, which was different from the CO generation from powdered activated carbon (PAC) thermolysis that initiated at the temperature over 720 °C. This substantial production of CO was attributed to redox reaction between Fe_3O_4 and CO_2. A small amount of Fe_3O_4 (Fe_3O_4/PVC = 0.01) was able to trigger a CO_2-looping gasification, increasing CO production in the temperature of 560-720 °C (up to 7.1 mole% at 680 °C). On the other hand, further addition of Fe_3O_4 at a ratio of 0.5 enhanced the concentration level of CO, particularly in the latter half of thermolysis (up to 4.5 mole % at 900 °C). Good adsorption capability of biochar from PVC/Fe_3O_4 thermolysis was also demonstrated to achieve over 77% removal efficiency of methylene blue.

Topics:
Carbon Management
Energy (Sustainability & Environment)

Utilization of Garment Thermosetting Plastic Button Residual Waste

- Conference: E2S2 CREATE and AIChE Waste Management Conference
- Year: 2019
- Proceeding: 2019 E2S2-CREATE and AIChE Waste Management Conference
- Group: Poster Submissions
- Session:
  Poster Session
- Time:
  Tuesday, March 12, 2019 - 3:55pm-4:20pm

Authors:
Hossain, M. I., Bangladesh University of Engineering and Technology

Garment industry has become the crucial component of the entire industrial sector in Bangladesh. It has an annual export value of approximately 38 billion U.S. dollars. Woven, knit and sweater garment items use buttons which are predominantly made of thermosetting plastic. The process of preparing those thermosetting plastic buttons always leaves residual plastic which is mostly discarded into environment as waste. Present study was aimed to determine promising routes for the utilization of that residual thermosetting plastic waste. The plastic waste was found to possess 6480 kcal/kg gross calorific value. It can therefore be considered to be used with solid fuel. X-ray fluorescence analysis revealed that the residual plastic waste contains 90 wt% carbon. The waste was subsequently processed into activated carbon which could successfully remove 90 wt% arsenic from contaminated water. The melted residual plastic waste was finally mixed with bituminous for the preparation of bituminous concrete. Marshall Stability test of the concrete explored that dosing the bituminous with residual thermosetting plastic waste tends to increase compressive strength of the concrete. It can therefore be concluded that the thermosetting plastic waste generated in garment industry has the potential to be utilized for better environmental and financial managements.

Topics:
Environmental Sustainability

AUTHOR INDEX

Ager, Joel W. 18
Ahamed, Ashiq 15
Ahmed, Shoeb 44
Alessi, Daniel 16
Ali, Syed M. Shamaim 40
Ashiq, Ahmed 33
Atugoda, Thilakshani 38
Badilla, Daisy 35
Barecka, Magda H. 18
Bolan, Nanthi 38
Cai, Yilu 37
Chan, Wei Ping 31
Chen, Hanping 9
Chen, Jialing 11
Chen, Xia 19
Chen, Yihan 56
Cho, Dong-Wan 61
Choi, Seung Wan 59
Chong, Clive 42
Chung, Jacky 24
Chung, Kyong-Hwan 49
Cui, Hongyou 19
Dai, Yanjun 11
Dou, Xiaomin 31
Ee, Alvin 47
Ee, Boon Siong 50
Fareed, Shamsoon 40
Fei, Xunchang 12
Francis, Tan 50
Galagedara, Heshan 29, 48
Geng, Yong 20
Giannis, Apostolos 15, 31
Gin, Karina Yew-Hoong 56
He, Yiliang 56
Hoque, Ayman 41
Hou, Deyi 8, 16
Hsu, S. C. 27
Hu, Chenlian 14, 60
Hu, Xiao 34
Huang, Yuansheng 56
Hwang, June H. 52, 53
Jayawardhana, Yohan 29, 45
Jayaweera, Supuli 34
Jeon, Ki-Joon 54
Jeong, Jaehun 22
Jiang, Jing 32
Jiang, Minmin 37
Jiang, Peng 37
Jo, Ji Hye 43
Jung, Sang-Chul 49
Kandasamy, Kishore 41
Khan, Asad-Ur-Rehman 40
Kim, Min-Jeong 55, 59
Kim, Min-Suk 52, 53
Kim, Woong 43
Klemes, Jirí Jaromír 36
Klemeš, J. Jaromír 2
Kua, Harn Wei 24
Kwon, Eilhann 21
Kwon, Gihoon 61
Lam, Su Shiung 13
Lee, Ki Bong 55, 59
Lee, Sang-Ryong 52, 53
Lei, Junxi 31
Li, Lanyu 42
Li, Peng 56
Li, Xian 11, 42
Lim, Jun Wei 30
Lim, Teik-Thye 31
Lima, Stephan Pflugmacher 25
Lin, Alexander 24
Lin, Jian 23
Lipinski, Wojciech 10
Lisak, Grzegorz 15, 31
Liu, Xiao 14, 32, 37, 60
Loh, Kai-Chee 28, 39, 58
Lu, Jie 5, 14, 60
Lu, Xuhong 12
Maag, Gilles 10
Mak, Tiffany M. W. 27
Mao, Liwei 39
Mathur, Pranjal 41
Mustafa, Maham 40
Ng, Wun Jern 34
Oh, Wen-Da 31
Ok, Yong Sik 26, 29, 33, 46, 53
Pan, Ze Hua 31
Pandey, Ashok 3
Pang, Patrick 1
Pang, Sze Dai 24
Park, Se Young 22
Park, Young Kwon 49
Park, Young-Kwon 22, 54
Poon, Chi Sun 27
Poyyamozhi, Sricharan 41
Pye, John 10

Quek, Augustine 47
Ramanayaka, Sammani 29
Ryu, Sumin 22
Sajid, Hamais 40
Sakib, Md. Nurus 44
Sarkar, Binoy 33
Sarvanantharajah, Suseeven 58
Seay, Jeffrey R. 6
Shen, Zhengtao 16
Sim, Kai Sheng 50
Snyder, Shane 4
Song, Hocheol 61
Tamilselvam, Kumaran S/o 31
Taylor, Hayden 24
Tian, Hailin 17
Tong, Yen Wah 17, 28, 30
Tsang, Daniel C. W. 7, 27
Van Fan, Yee 2, 36
Veksha, Andrei 15, 31
Venkataraman, Mahesh 10
Vithanage, Meththika 29, 33, 38, 45, 48
Wang, Chi-Hwa 11, 24, 42
Wang, Ming 19
Wang, Xiaonan 17, 26, 42, 46
Weerachanchai, Piyarat 15
Weerasooriyagedara, Madara 33
Wijayasinghe, Athula 29, 48
Wijesekara, Hasintha 38
Xu, Li-Hua 57
Yang, Haiping 9
Yin, Ke 15, 34
Yu, Iris K. M. 27
Yu, Kaifeng 56
Yuan, Xiangzhou 55
Zhan, Xiangyan 23
Zhang, Bo 56
Zhang, Jingxin 28, 39
Zhang, Le 39, 58
Zhang, Yali 19
Zhao, Yingru 23
Zheng, Xuyue 23
Zhu, Xingyi 23
Zhu, Xinzhe 26, 46

AIChE
120 Wall Street, FL 23
New York, NY 10005-4020

ISBN 978-1-7138-0578-6

proceedings.com